A Short Guide to Writing about Biology

A Short Guide to Writing about Biology

Third Edition

Jan A. Pechenik
Tufts University

An imprint of Addison Wesley Longman, Inc.

New York • Reading, Massachusetts • Menlo Park, California • Harlow, England
Don Mills, Ontario • Sydney • Mexico City • Madrid • Amsterdam

Acquisitions Editor: Patricia Rossi
Project Coordination and Text Design: York Production Services
Supervising Production Editor: Lois Lombardo
Cover Design: Scott Russo
Manufacturing Manager: Hilda Koparanian
Electronic Page Makeup: ComCom
Printer and Binder: RR Donnelley & Sons Company
Cover Printer: The Lehigh Press, Inc.

Library of Congress Cataloging-in-Publication Data
Pechenik, Jan A.
A short guide to writing about biology / Jan A. Pechenik. — 3rd ed.
p. cm.
Includes bibliographical references (p.) and index.
ISBN 0-673-52503-1
1. Biology—Authorship. 2. Report writing. I. Title.
QH304.P43 1996 96-35994
CIP

ISBN 0-673-52503-1
12345678910—DOC—99989796

To Oliver,
my f.l.g.

Contents

Preface

Communicating through writing is an important part of the biologist's trade, and many of us wish we had more time to teach our students to do it well. Bad scientific writing often reflects fuzzy thinking, so that questioning the writing generally guides students toward a clearer understanding of the biology being written about. To improve students' writing in biology, half the battle is to persuade them to take the task seriously; the other half is to provide them with sufficient instructions to get them through the struggle successfully. In this book, I address both halves of the battle in a way that should not take up valuable class time.

The book is brief enough to be read by students along with other, more standard assignments, and straightforward enough to be understood without additional instruction. Although intended primarily for undergraduate use in typical lecture and laboratory courses at all levels, the book is also appropriate for undergraduate and graduate seminars. Many colleagues also tell me that they have found much in the book that was new and helpful.

Throughout the book, I emphasize that careful thinking cannot be separated from effective writing. With the dramatic innovations in molecular techniques and information technology, students can easily lose sight of what Biology—and being a biologist—is really all about. It is not really about memorizing esoteric terminology and mastering an increasing array of computer software and molecular techniques. Biology is a way of thinking about the world. It is about making careful observations, asking specific questions, designing ways of addressing those questions, manipulating data thoughtfully and thoroughly, interpreting those data and related observations, reevaluating past work, asking new questions, and redefining older ones. It is also about communicating information—accurately, logically, clearly, concisely, and unambiguously. The hard work of thinking about Biology is at least as important as the work of doing it, and writing provides a way to examine, to evaluate, and to refine that thinking.

Organization

Chapters 1, 2, 4, and 10 should be read by all students early in the term before beginning any writing assignments; the other chapters cover the specific writing assignments typically encountered in the Biology undergraduate curriculum and can be assigned as appropriate, in any order.

Chapter 1 presents the benefits of learning to write well in Biology; it also describes the sort of writing that professional biologists do, and reviews the key principles that characterize all sound scientific writing. This chapter also includes a section on the use of computers in searching the literature and preparing writing assignments. Chapter 2 gives detailed instruction about how to locate research articles using various printed and computerized indexing services, how to read the formal scientific literature—including graphs and tables—and how to take useful notes; it emphasizes the struggle for understanding that must precede any concern with *how* something is said. Chapter 4 discusses the writing of summaries and critiques. Most students would benefit substantially from additional practice in summarizing what they read and hear; successful evaluation and synthesis depend on the ability to summarize.

Chapter 10 focuses on the process of revision and emphasizes the usefulness of peer review; it includes a short section on how to be an effective reviewer of other people's writing and how to interpret the criticism received.

Chapter 3, "Writing Laboratory Reports," emphasizes that

the result obtained in a study is often less important than the ability to discuss and interpret that result convincingly in the context of basic biological knowledge, and to demonstrate clear understanding of the purpose of the study. I emphasize the variability inherent in biological systems and how that variability is dealt with in presenting, interpreting, and discussing data. This chapter should also be useful for students preparing papers for publication. Chapter 5, "Writing Essays and Term Papers," discusses the most profitable ways to decide on and explore a paper topic.

Writing research proposals and in-class essay examinations are discussed in separate chapters (Chapters 6 and 11). There is an additional chapter (Chapter 7) on writing for general nonscience audiences. To make the book more useful to all Biology majors, I have included chapters on how to give oral presentations (Chapter 8), how to prepare applications for jobs in Biology and for graduate and professional schools (Chapter 9), and how to create effective poster presentations (Chapter 12). There is also an appendix listing commonly used abbreviations for lengths, weights, volumes, and concentrations, and another appendix listing sources of computer software that is especially useful to biologists.

CHANGES MADE FOR THE THIRD EDITION

I have added two new chapters to the book. Chapter 12 gives suggestions on preparing effective poster presentations; posters typically follow the precise format and layout of journal presentations but should not do so. This chapter is based partly on my own experience viewing posters at meetings and partly on a poster presentation given by Professor John D. Woolsey (Moore College of Art and Design) at the 1994 meeting of the American Society of Zoologists (now the Society for Integrative and Comparative Biology). Chapter 7 discusses the art of science journalism, not because I think that we should be preparing our students for careers as reporters but because such assignments turn out to be excellent ways to help students dejargonize their writing, and to help instructors tell just how much their students understand about the material they are writing about.

I have expanded the chapter on writing summaries and critiques and placed it earlier in the book to emphasize the importance of learning how to summarize information accurately and concisely. This is not something that most students seem to have had much practice doing, and that is a serious obstacle to both synthesis and critical evaluation. Writing summaries is also a particularly effective way for students to self-test their understanding.

I have added a few new rules to Chapter 1, and now discuss how to make effective slides and overheads in the chapter on oral presentations (Chapter 8). Chapter 2 now includes additional information on searching computer databases and prowling the Web. The chapter on writing laboratory reports and research papers (Chapter 3) now includes advice on making laboratory and field drawings, and more ideas about ways of plotting variability about means. The section on proper use of words and numerals when writing about numbers has been updated to reflect current CBE (Council of Biology Editors) recommendations. I have also summarized at the end of Chapter 3 the rules to be followed in writing each section of a research report. Students can use this checklist to assess their reports before submitting them and to peer-review drafts written by fellow students.

Users of the third edition will notice many other, smaller improvements throughout the book. The book has benefited from the advice and suggestions of the many people who took the time to read and comment on the two previous editions and on the manuscript for the present edition: Sylvan Barnet; Anne Kozak, College of the Atlantic; John R. Diehl, Clemson University; Tina Ayers, Northern Arizona University; David Takacs, Cornell University; Stephen Van Scoyoc, University of Delaware; Barbara

Stewart, Swarthmore College; and Carl Schaefer, University of Connecticut. It is also a pleasure to thank my colleagues Sarah Lewis, for helpful conversations about statistics and graphs, and Norman Hecht, for being such a good (and patient) sport during my work on Chapter 7. I have learned much about writing and teaching from letters and conversations with the many enthusiastic users of the book, and I welcome additional comments from readers of the present edition.

Jan A. Pechenik
jpecheni@emerald.tufts.edu

What appears as a thoroughly systematic piece of scientific work is actually the final product: a cleanly washed offspring that tells us very little about the chaotic mess that fermented in the mental womb of its creator.

Auner Treinin

Beginnings are hardest. . . . How can you write the beginning of something till you know what it's the beginning of?

Peter Elbow

Unlike other forms of writing, such as fiction where the goals are impossible to define, scientific writing has a single objective: to inform the reader.

Michael Alley

The truth is that badly written papers are most often written by people who are not clear in their own minds what they want to say.

John Maddox

Write to illuminate, not to impress.

Attributed to the biologist Naj A. Kînehcép

1
Introduction and General Rules

The logical development of ideas and the clear, precise, and succinct communication of those ideas through writing are among the most difficult, but most important, skills that can be mastered in college. Effective writing is also one of the most difficult skills to teach. This is especially true in Biology classes, where there is often much writing to be done but little time to focus on doing it well. The chief message of this book is that developing your writing skills is worth every bit of effort it takes, and that Biology is a splendid field in which to pursue this goal.

WHAT DO BIOLOGISTS WRITE ABOUT, AND WHY?

The sort of writing that biologists do (lectures, letters of recommendation, grant proposals, research papers, and critiques of research papers and grant proposals written by other researchers) is similar in many respects to the sort of writing (essays, literature reviews, term papers, and laboratory reports) you are asked to do while enrolled in a typical Biology course. Basically, we must all prepare arguments.

Like a good term paper or essay, a lecture is an argument; it presents information in an orderly manner and seeks to convince the audience that this information fits sensibly into some much larger story. Few students are aware of the time and effort required to write a coherent lecture, but the sad fact is that putting together a string of three or four lectures before moving on to the next topic is the equivalent of preparing one 20- to 30-page term paper weekly.

In addition to preparing lectures, many of us spend quite a bit of time writing grant proposals in the hope of obtaining the funding that will enable us to pursue our research programs (and possibly hire one of you for a summer in the process). A research proposal is unquestionably an argument; success depends on our ability to convince a panel of other biologists that what we wish to do is worth doing, that we are capable of doing it, that we are capable of correctly interpreting the results, and that the work cannot be done without the funds requested. Research money is not plentiful. Even well-written proposals have a difficult time; poorly written proposals generally don't stand a chance.

When we are not writing grant proposals or lectures, we are often preparing the results of our research for publication. Essentially, these articles are laboratory reports based on data collected over a much longer period than the typical laboratory session; in research articles, as in the preparation of laboratory reports, the goal is to present data clearly and to interpret those data thoroughly and convincingly in the context of other work and basic biological principles. The preparation of research reports typically involves the following steps:

Organizing the data

Preparing a first draft of the article (following the procedures outlined in Chapter 3 of this book)

Revising and reprinting (or retyping) the paper

Asking one or several colleagues to read the paper critically

Revising the paper in accordance with the comments and suggestions of the readers

Reprinting (or retyping) and proofreading the paper

Sending the paper to the editor of the journal in which we would most like to see our work published

This is not the end of the story. The editor then sends the manuscript out to be reviewed by two or three other biologists. Their comments, along with those of the editor, are then sent to the author, who must again rewrite the paper, often extensively. The editor may then accept or reject the revised manuscript, or may request that it be rewritten again prior to publication.

Biologists obviously write about Biology, but they also write about other things. One of the other things college and university biologists write about is you; letters of recommendation are especially troublesome for us because they are so important to you. Like a good laboratory report, literature review, essay, or term paper, a letter of recommendation must be written clearly, developed logically, and proofread carefully if it is to argue convincingly on your behalf and help get you where you want to go.

And then there are the progress reports, committee reports, and internal memoranda. All this writing involves thinking, organizing, nailing down a convincing argument on paper, revising, retyping, and proofreading.

I hope you are now convinced that effective writing is not irrelevant in a scientific career. When students in a Biology course receive criticisms of their writing, they often complain that "this is not an English course." These students do not understand that clear, concise, logical writing is an important tool of the biologist's trade, and that learning how to write well is at least as important as learning how to use a balance, extract a protein, use a taxonomic key, measure a nerve impulse, or run an electrophoretic gel. And, unlike these rather specialized laboratory techniques, mastering the art of effective writing will reward you regardless of the field in which you eventually find yourself. The fact that you may not become a biologist is no reason to cheat yourself out of the opportunity to become an effective writer; the difference between a well-crafted and poorly crafted letter of application is often the difference between getting the job you want or losing to another contender.

THE KEYS TO SUCCESS

There is no easy way to learn to write well in Biology. All good writing involves two struggles: the struggle for understanding and the struggle to communicate that understanding to a reader. Like the making of omelettes or crepes, the skill improves with practice. However, being aware of certain key principles will ease the way considerably. Each of the following rules is discussed more fully in later chapters.

1. **Work to understand your sources.** When writing laboratory reports, spend time wrestling with your data until you are convinced you see the significance of what you have done. When taking notes from books or research articles, reread sentences you don't understand and look up the words that puzzle you. Try to take notes in your own words; extensive copying or paraphrasing usually means that you do not yet understand the material well enough to be writing about it. Too few students take this struggle for understanding seriously enough, but all good scientific writing begins here. You can excel—in college and in life after college—by being one of the few who meet this challenge head on. Do not be embarrassed to admit (to yourself or to others) that you do not understand something after working at it for a while. Talk about the problem with other students or with your instructor. If you don't commit yourself to winning the struggle for understanding, you will end up with nothing to say, or worse, what you do say will be

wrong. In both cases, you will produce nothing worth reading.

2. **Think about where you are going before you begin to write.** Much of the real work of writing is in the thinking that must precede each draft. Effective writing is like effective sailing; you must take the time to plot your course before getting too far from port. Your ideas about where you are going and how best to get there may very well change as you continue to work with and revise your paper, since the act of writing invariably clarifies your thinking and often brings entirely new ideas into focus. Nevertheless, you must have some plan in mind even when you begin to write your first draft. This plan evolves from thoughtful consideration of your notes. Think first, then write; thoughtful revision follows. Some people find it helpful to think at a keyboard or with pen in hand, letting their thoughts tumble onto paper. Others prefer to think "inside," writing only after their thoughts have come together into a coherent pattern. Either way, the hard work of thinking must not be avoided. If, when you sit down to write that last draft of your paper, you still don't know where you are heading, you certainly won't get there smoothly and you may well not get there at all. Almost certainly, your readers will never get there.
3. **Allow time for revision.** Accurate, concise, successfully persuasive communication is not easily achieved, and few of us come close in a first or even a second draft. Although the act of writing can itself help clarify your thinking, it is extremely helpful to step away from the work and reread it with a fresh eye before making revisions; a "revision" is, after all, a re-vision: another look at what you have written. This second (or third, or fourth) look allows you more easily to see if you *have* said what you had hoped to say, and whether you have guided the

reader from point to point as masterfully as you had intended. Start writing assignments as soon as possible after receiving them, and always allow at least a few days between the penultimate and final drafts; if you follow this advice, the quality of what you submit will improve dramatically, as will the quality of what you learn from the assignment.

4. **Write to illuminate, not to impress.** Use the simplest words and the simplest phrasing consistent with that goal. Avoid acronyms, and define all specialized terminology. In general, if a term was recently new to you, it should be defined in your writing. And if you can talk about "zones of polarizing activity" instead of "ZPAs," please do so. Your goal should be to communicate; why deliberately exclude potentially interested readers by trying to sound "scientific"? Don't try to impress the reader with big words and a technical vocabulary; focus instead on getting your point across.
5. **Make a statement and back it up.** Remember, you are making an argument. In any argument, a statement of fact or opinion becomes convincing to the critical reader only when that statement is supported by evidence or explanation; provide it. You might, for instance, write, "Among the vertebrates, the development of sperm is triggered by the release of the hormone testosterone (Gilbert, 1991)." In this case, the statement is supported by reference to a book written by Scott Gilbert in 1991. In the following example, a statement is backed up by reference to the writer's own data: "Some wavelengths of light were more effective than others in promoting photosynthesis. For example, the rate of oxygen production at 650 nm* was nearly four times greater than that

*nm = nanometers; that is, 10^{-9} meters.

recorded for the same plants when using a wavelength of 550 nm (Figure 2)."

References to papers or books written by two authors must include the names of both authors (for example, Burns and Allen, 1946). When there are more than two authors, only the first author's name is written out (for example, Fried et al., 1995). Note that an author's first name is never included in the citation. A statement made by your instructor should be cited as a personal communication (for example, "Professor Rachel Merz, personal communication"). Refer to a laboratory manual or handout by the author of the handout (for example, R. Chase, 1996), or as follows: (Laboratory Manual, 1996).

6. **Always distinguish fact from possibility.** In the course of examining your data or reading your notes, you may form an opinion. This is splendid. But you must be careful not to state your opinion as though it were fact. "Species *X* lacks the ability to respond to sucrose" is a statement of fact and should be supported with a reference. "Our data suggest that species *X* lacks the ability to respond to sucrose" or "Species *X* seems unable to respond to sucrose" expresses your opinion and should be supported by drawing the reader's attention to key elements of your data set.
7. **Say exactly what you mean.** Words are tricky; if they don't end up in the right places, they can add considerable ambiguity to your sentences. "I saw three squid SCUBA diving last Thursday" conjures up a very interesting image. Don't make readers guess what you're trying to say; they often guess incorrectly. Good scientific writing is precise. Write to mean what you mean to say, and be sure you say what you mean. It often helps to read aloud what you have written and to listen to what you are saying as you read.

8. **Never make the reader back up.** You should try to take the reader by the nose in your first paragraph and lead him or her through to the end, line by line, paragraph by paragraph. Avoid making the reader flip back two pages, or even one sentence. Link your sentences carefully, using such transitional words as "Therefore" or "In contrast," or by repeating key words, so that a clear argument is developed logically. Remind the reader of what has come before, as in the following example:

    In saturated air (100% relative humidity), the worms lost about 20% of their initial body weight during the first 20 hours but were then able to prevent further dehydration. In contrast, worms maintained in air of 70-80% relative humidity experienced a much faster and continuous rate of dehydration, losing 63% of their total body water content in 24 hours. As a consequence of this rapid dehydration, most worms died within the 24-hour period.

    Note that the second and third sentences in this example begin with transitions ("In contrast," "As a consequence of"), thus continuing and developing the thought initiated in the preceding sentences. A far less satisfactory last sentence might read, "Most of these animals died within the 24-hour period."

    Link your paragraphs in the same way, using transitions to continue the progression of a thought, reminding the readers periodically of what they have already read.

    Avoid casual use of the words *it, they,* and *their.* For example, the sentence "It can be altered by several environmental factors" forces the reader to go back to the preceding sentence, or perhaps even to the previous paragraph, to find out what *it* is. Changing the sentence to "The rate of population growth can be altered by several environmental factors" solves the problem. Here is another example:

```
Our results were based upon observations of
short-term changes in behavior. They showed that
feeding rates did not vary with the size of the
caterpillar.
```

The word *they* could refer to "results," "observations," or "changes in behavior." Granted, the reader can back up and figure out what *they* are, but you should work to avoid the "You know what I mean" syndrome. Changing *they* in the second sentence to "These results" avoids the ambiguity and keeps the reader moving in the right direction.

Do not be afraid to repeat a word used in a preceding sentence; if it is the right word and avoids ambiguity, use it.

9. **Don't make readers work harder than they have to.** If there is interpreting to be done, you must be the one to do it. For example, never write something like

```
The difference in absorption rates is quite
clearly shown in Table 1.
```

Such a statement puts the burden of effort on the reader. Instead, write something like

```
Clearly, alcohol is more readily absorbed into
the bloodstream from distilled, rather than
brewed, beverages (Table 1).
```

The reader now knows exactly what you have in mind and can examine Table 1 to see if he or she agrees with you.

10. **Be concise.** Give all the necessary information, but avoid using more words than you need for the job at hand. By being concise, your writing will gain in clarity. Why say,

    `Our results were based upon observations of short-term changes in behavior. These results showed that feeding rates did not vary with the size of the caterpillar.`

    when you can say

    `Our observations of short-term changes in behavior indicate that feeding rates did not vary with the size of the caterpillar.`

    In fact, you might be even better off with the following sentence:

    `Feeding rates did not appear to vary with the size of the caterpillar.`

    With this modified sentence, 50 percent of the words in the first effort have been eliminated without any loss of content. The savings are not merely esthetic. It costs something like 20 cents a word to publish a scientific paper, and authors are often asked to bear some of this cost; in the real world of biological publications, it pays, quite literally, to be concise. Besides, cutting out extra words means you will have less to type. You'll have your paper finished that much sooner. Finally, your readers can digest the paper more easily, reading it with pleasure rather than with impatience.

11. **Stick to the point.** Delete any irrelevant information, no matter how interesting it is to you. Snip it out and put it away in a safe place for later use if you wish, but don't let asides interrupt the flow of your writing.
12. **Write for your classmates and for your future self.** It is difficult to write effectively unless you have a suitable audience in mind. It helps to write papers that you can imagine being understood by your fellow students. You should also prepare your assignments so that they will be meaningful to *you* should you read them far in the future, long after you have forgotten the details of coursework completed or experiments performed. Addressing these two audiences—your fellow students and your future self—should help you write clearly and convincingly.
13. **Don't plagiarize.** Express your own thoughts in your own words. Quotations are rarely used in the biological literature; if you are quoting something simply because you don't understand what you are reading, try to figure it out on your own or ask for help. If you do quote from another writer or restate that writer's ideas or interpretations, you must credit your source explicitly. Note, too, that simply changing a few words here and there or changing the order of a few words in a sentence or paragraph is still plagiarism. Plagiarism is one of the most serious crimes in academia: it can get you expelled from college or cost you a career later. With practice and conscientious effort, you will find yourself capable of generating your own good ideas and presenting them in perfectly fine prose of your own devising.
14. **Don't be teleological.** That is, don't attribute a sense of purpose to other living things, especially when discussing evolution. Giraffes did not evolve long necks "in order to reach the leaves of tall trees." Snails did not

evolve shells "in order to confound predators." Birds did not evolve nest-building behavior "in order to protect their young." Insects did not evolve wings "in order to fly." Plants did not evolve flowers "in order to attract bees for pollination." Natural selection operates through a process of differential survival and reproduction, not with intent. Long necks, hard shells, complex behavior, and other such genetically determined characteristics may well have given some organisms an advantage in surviving and reproducing unavailable to individuals lacking these traits, but this does not mean that any of these characteristics were deliberately evolved in order to achieve something.

Organisms cannot evolve structures, physiological adaptations, or behavior out of desire. Appropriate genetic combinations must always arise by random genetic events, by chance, before selection can operate. Even then, selection is imposed on the individual by its surroundings and, in that sense, selection is a passive process; natural selection never involves conscious, deliberate choice. Don't write, "Insects may have evolved flight in order to escape predators." Instead, write, "Flight in insects may have been selected for in response to predation pressure." Don't write, "The parent gulls remove the white, conspicuous eggshells in order to protect the newly hatched, black-headed young." Instead, write, "Parental removal of the white, conspicuous eggshells may protect the newly hatched, black-headed young from predation."

15. **Always underline or italicize species names as in, for example, *Homo sapiens*.** Note also that the generic name (*Homo*) is capitalized, whereas the specific name (*sapiens*) is not. Once you have given the full name of the organism in your paper, the generic name can be abbreviated; *Homo sapiens,* for example, becomes *H. sapiens.*

There is no other acceptable way to abbreviate species names. In particular, it is not permissible to refer to an animal using only the generic name, since most genera include many species. Note that the plural of "genus" is "genera," not "genuses."

16. **Capitalize the names of taxonomic groups (clades) above the level of genus, but not the names of the taxonomic categories themselves.** For example, insects belong to the phylum Arthropoda and the class Insecta. Do not capitalize informal names of animals: insects are arthropods, members of the phylum Arthropoda.
17. **Remember that the word *data* is plural.** The singular is *datum,* a word rarely used in biological writing. "The data are lovely" (not "The data is lovely"). "These data show some interesting trends" (not "This data shows some interesting trends"). You would not say "My feet is very large"; treat *data* with the same respect.
18. **Proofread.** Although it is an important part of the writing process, none of us likes to proofread. By the time we have arrived at this point in the project, we have put in a considerable amount of work and are certain we have done the job correctly. Who wants to read the paper yet another time? Moreover, finding an error means having to make a correction. But put yourself in the position of your instructor. Your instructor must read perhaps a hundred or more papers each term. He or she starts off on your side, wanting to see you earn a good grade. Similarly, a reviewer or editor of scientific research manuscripts starts off by wanting to see the paper under consideration get published. A sloppy paper—for example, one with many typographical errors—can lose you a considerable amount of good will as a student and later as a practicing scientist. For one thing, sloppy work may suggest to the reader that you are equally sloppy in your thinking, or that you take little pride in your own efforts. Furthermore, it's insulting. Fail-

ure to proofread your paper and to make the required corrections implies that you don't value the reader's time; that is not a flattering message to send, nor is it a particularly wise one. Never forget: there is often a subjective element to grading and to decisions about the fate of manuscripts and grant proposals. Lastly, a carelessly proofread paper may suggest to the reader that the research itself was carelessly performed. For all these reasons, shoddily prepared material can easily lower a grade, damage a writer's credibility, reduce the likelihood that a manuscript will be accepted for publication or that a grant proposal will be funded, or cost an applicant a job or admission to professional or graduate school. Why put yourself in such jeopardy for a mere half-hour saved? Turn in a piece of work that you are proud to have produced.

19. **Remember that appearances can be deceiving.** Your papers and reports should give the impression that you took the assignment seriously, that you are proud of the result, and that you welcome constructive criticism of your work. Type or computer-print your papers if possible, using only one side of each page. Leave margins of about an inch and a half on the left and right sides of the page, leave about an inch at the top and bottom of each page, and double-space your typing so that your instructor can easily make comments on your paper. Make corrections neatly. Never underestimate the subjective element in grading.
20. **Put your name and the date at the top of each assignment, and number all pages.** Pages should be numbered so that the reader can tell immediately if a page is missing or out of order and so that the reader can easily point out problems on particular pages ("In the middle of page 7, you imply that . . .").
21. **Always make a copy of your paper before submitting the original to your instructor.** Even instructors sometimes lose things.

ON USING COMPUTERS IN WRITING

Computers are a writer's best friend when it comes to revising advanced drafts of manuscripts and reports, and it is here that you can exploit those disk drives to best advantage. Back in the days of typewriters, when reading drafts of my papers I would often see places where rearranging a few paragraphs, adding a phrase or sentence, or even simply replacing one word with another, would have substantially improved the final product. But if I had already typed several drafts, I rarely made those additional changes; the benefits of increased clarity of expression were usually overwhelmed by the unbearable thought of retyping one or more pages yet again. With word processing, however, perfection is within your immediate grasp. It is now easy to change a word, modify or delete a sentence, or reorganize a paragraph or an entire paper, and the computer will produce the revised version at the touch of a button.

Even so, I encourage you to write your first drafts with pencil and paper. First drafts serve primarily to get ideas on paper, where they can't escape; the form, order, and manner of expression are not major concerns at this early stage of creation. Consequently, the first revision is often so extensive that it is far less time consuming to revise this draft by hand than to do so with a computer. In fact, putting a first draft on the computer might actually inhibit you from making the extensive revisions that are called for. Moreover, when there is a power failure, you can lose everything; if you use a pencil or pen instead, you will still have your first, handwritten draft. For me, it's the second draft that gets entered into the computer.

If you are one of those people who are intimidated by a blank sheet of paper but not by a blank video screen, ignore my advice about not using a computer for first drafts. Otherwise, reserve the computer for a later version of your assignment.

Let me end this section with some warnings about what you cannot expect a computer to do for you. Advertisers suggest that owning a computer will take the work out of your writing.

Armed with a personal computer, they say, you will see your spelling improve, your sentences make sense, your paragraphs become well organized, your ideas seem brilliant, and your grades soar. As the ever-optimistic reader of many computer-printed laboratory reports and papers, I must inform you that computers do not work these kinds of miracles. Word processing programs, unfortunately, cannot remove from the author the responsibility for thinking, organizing, revising, and proofreading. In particular, they can do little to help you in the first struggle—the struggle for understanding—and they cannot think, organize, or revise for you. Computerized spelling checkers are of some use in catching your typographical and spelling errors, but you cannot expect them to catch all of your mistakes. Biology is a field with much specialized terminology, much of which is of no use to nonbiologists; these terms, therefore, do not find their way into the dictionaries that accompany computerized spelling programs. Although you can easily add words to the computer's dictionary, the terminology in your papers will be changing with every new assignment; many of the words you add for today's assignment will probably not be used in next week's assignment. Moreover, a spelling-checker program will not distinguish between *to* and *too, there* and *their,* or *it's* and *its,* and the program will miss typographical errors that are real words; using the program will not spare you the chore of proofreading for spelling mistakes. Suppose, for example, that you typed *an* when you intended to type *and,* or you typed *or* when you should have typed *of,* or you typed *rat* instead of *rate.* To catch these errors, you would have to use a program that catches grammatical mistakes, but be aware that such programs will not catch every error and do not always suggest the proper correction when mistakes are recognized. By all means use spelling- and grammar-checking programs if they are readily available, but then use your own sharp eyes and keen intellect—moving word by word and sentence by sentence—to complete the necessary process of proofreading your work.

Be aware that word processing is a two-edged sword; because revisions no longer require time-consuming retyping, use of a computer places increased responsibility on the writer to see that the revisions get made. Instructors find it increasingly annoying when students turn in computer-printed reports that are carelessly written and not proofread.

ON USING COMPUTERS FOR DATA STORAGE, ANALYSIS, AND PRESENTATION

In addition to their use as word processors, computers are also used by many biologists for storing and retrieving literature references and for storing and analyzing especially large or complex data sets. But undergraduate Biology majors will probably find that a set of notecards and a $30 scientific calculator will be perfectly adequate for anything they will be asked to do in most courses.

On the other hand, computers can be a real help in preparing your graphs and tables, as discussed in Chapter 3 (p. 73). Even so, you should not feel compelled to generate your figures and tables by computer; unless told otherwise by your instructor, hand-drawn graphs and tables should earn you as good a grade, provided they are carefully planned, sensible, and neatly executed.

SUMMARY

1. Acknowledge the struggle for understanding, and work to emerge victorious; read with a critical, questioning eye.
2. Think about where you are going before you begin to write, while you write, and while you revise.
3. Allow adequate time for revision.
4. Write to illuminate, not to impress.
5. Back up all statements of fact or opinion.
6. Always distinguish fact from possibility.

7. Say exactly what you mean.
8. Never make the reader back up.
9. Don't make the reader work harder than he or she needs to.
10. Be concise: avoid unnecessary words, weak verbs, and unnecessary prepositions.
11. Stick to the point.
12. Write for an appropriate audience: your classmates and your future self.
13. Don't plagiarize.
14. Avoid teleology.
15. Underline or italicize the scientific names of species.
16. Remember the word *data* is plural, not singular.
17. Proofread all work before turning it in, and keep a copy for yourself.
18. Make your papers neat in appearance, double-space all work, and leave margins for the instructor's comments and suggestions.
19. Put your name and the date at the top of each assignment, and number all pages.
20. When possible, use word processing computer programs for revising advanced drafts of manuscripts, but prepare the first draft or two with pencil and paper.

2
General Advice on Reading and Note-Taking

EFFECTIVE READING

Too many students think of reading as the mechanical act of moving the eyes left to right, line by line, to the end of a page, and repeating the process page after page to the end of a chapter or an assignment. I call this "brain-off" reading. When the last page has been "read," the task is over and it's on to something else. This is, after all, the way we typically watch television: We sit transfixed before the television until the program has ended, and then either change the channel or turn the set off; we've "seen" the program. In the same way, students typically "listen" to a lecture by furiously copying whatever the instructor writes or says, without really thinking about the information being presented. However, if you hope to develop something worth saying in your writing, you must *interact* intellectually with the material; you must become a "brain-on" reader, wrestling thoughtfully with every sentence, every graph, every illustration, and every table. If you don't fully understand any element of what you are reading (including your lecture notes), you must work through the problem until it is resolved rather than skipping over the difficult material and moving along to something more accessible.

This is inevitably a time-consuming process, but you can do a number of things to smooth the way. Whether you are writing an essay, a term paper, or the introduction or discussion section of a laboratory report or research article, always begin by carefully reading the appropriate sections of your textbook and class notes to get a solid overview of the general subject of which your topic is a part. It is usually wise to then consult one or two additional textbooks before venturing into the "primary literature" that reports the results of original research; a solid construction requires a firm foundation. Your instructor may have placed a number of pertinent textbooks on reserve in your college library. Alternatively, you can consult the library filing system, looking for books listed under the topic you are investigating. I will say more about locating appropriate sources at the end of this chapter.

Armed with this background information, you are now prepared to delve into more advanced textbooks and the primary scientific literature.

Reading a formal scientific paper is unlike reading a work of fiction or even a textbook or review article. The primary scientific literature must be read slowly, thoughtfully, and patiently, and a single paper must usually be reread several times before it can be thoroughly understood; don't become discouraged after only one or two readings. Reading the scientific literature is slow going, but, like playing tennis or sight-reading music, it gets easier with practice. If, after several rereadings of the paper, and if, after consulting several textbooks, you are still baffled by something in the paper you are reading, ask your instructor for help.

As you carefully read each paper, pay special attention to the following:

1. What specific question is being asked?
2. How does the design of the study address the question posed?
3. What are the controls for each experiment?
4. How convincing are the results? Are any of the results surprising?

5. What contribution does this study make toward answering the original question?
6. What aspects of the original question remain unanswered?

READING DATA

Data—the most important parts of any book or journal article—are displayed either as figures or as tables, and it is important to develop the skills needed to examine these elements critically. Your goal here is to come to your own interpretation of the data so that you can better understand or evaluate the author's interpretation. To do this, you must study the data and ask yourself some questions about how the study was done, why it was done, and what the major findings were.

Consider the example shown in Figure 2.1, modified from a 1990 review paper entitled "Peptide Regulation of Mast-Cell Function," by David E. Cochrane. From your background reading or class lecture notes, you would probably know that mast cells release into the blood a variety of substances involved in provoking allergic and inflammatory responses. If you didn't already know this, you would do some background reading in your textbook before proceeding.

Looking at Figure 2.1, let us see if we can figure out what the researchers did to collect their data. By reading the axis labels, we learn that the graph shows how blood histamine concentrations change over time, and we learn from the figure caption that these changes are provoked by a particular peptide called neurotensin (NT). The study was done on anesthetized rats, and all the action seems to occur rather quickly since the X-axis extends only to 30 minutes. Looking more closely, we see that histamine concentrations were initially quite low (less than 1 nanogram* per 10 microliters** of blood plasma), as indicated by the arrow (at

*1 nanogram (ng) = 10^{-9} gram.
**1 microliter (μl) = 10^{-6} liter.

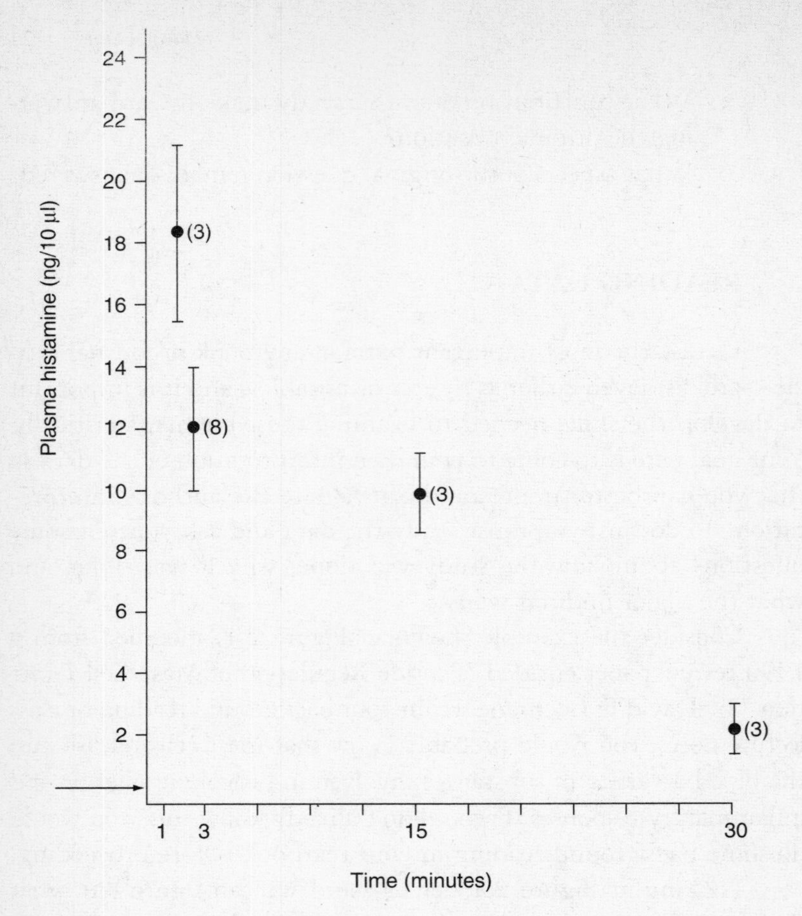

Figure 2.1
Plasma histamine concentrations in response to neurotensin (NT) given at t = 0. Rats were anesthetized and given NT (5 nmol/kg) or saline (0.3 ml) intravenously. Blood samples were collected at the indicated times. Each point represents the mean (± one standard error about the mean) of n values (given in parentheses). The horizontal arrow (lower left) shows the mean histamine concentration before addition of NT. Intravenous injection of saline did not alter this concentration over the 30-minute observation period. From D. E. Cochrane, 1990. Peptide regulation of mast-cell function. In, *Progr. Medicinal Chemistry,* Vol. 27; G. P. Ellis and G. B. West, eds.; Elsevier Science Publ. (Biomedical Division), pp. 143–188.

the lower left side of the graph), and that they rose impressively by the time the first blood sample was taken one minute into the study. Other blood samples were taken 3, 15, and 30 minutes after neurotensin was administered, and three to eight separate samples were taken at each time period. Even without reading the figure caption and without seeing the numbers alongside each data point, we would know that replicate samples were taken, since the thin lines extending vertically from each point indicate the amount of variation seen about the mean value recorded for each sampling time; obviously one can see variation about a mean value only when multiple samples are taken.

What were the controls for this experiment? Apparently a number of rats were injected with a saline solution instead of with neurotensin, and blood samples were also taken from these rats at appropriate times.

So, we know quite a bit about how this aspect of the study was conducted just by scrutinizing the graph. It helps considerably that the graph and figure caption were carefully constructed. What additional information might you wish to have? Here are a few questions that might arise as you think about the figure.

1. How did Dr. Cochrane decide to inject NT at a concentration of 5 nmol/kg? Is this a physiologically realistic concentration?
2. How was histamine concentration determined?
3. How many rats served as controls for each time interval?
4. How old were the rats? What sex were they? Where were they obtained from?
5. How were the rats anesthetized, and might this pretreatment affect the response to NT?
6. With regard to the replicate samples, were three to eight blood samples taken simultaneously from a single rat at each sampling time, or were blood samples taken from three to eight different rats each time? Blood samples were *probably* taken from a number of

different individual rats, but we can't be sure from the graph.

7. Were different rats sampled at each time interval, or were the same individuals sampled repeatedly, for example, at 15 minutes and again at 30 minutes? One might guess that the same individuals were sampled repeatedly. On the other hand, there may be technical limitations to drawing blood from any individual rat more than once. Suppose, for instance, that a very large volume of blood was required to determine the histamine concentration. This issue leads to my last two questions.
8. What volume of blood was withdrawn for each sample?
9. How were the blood samples obtained?

Asking such questions puts you in a good position to interpret the results illustrated and provides a framework for your later reading of the Materials and Methods section of the report; you can now enter the methodology section of the paper looking for the answers to specific questions.

If asked to describe how this aspect of the study was conducted and why the study was probably undertaken, how might you respond? Even without reading anything else in the paper, you could write the following summary:

> This study was apparently undertaken to determine the ability of the peptide neurotensin to elicit histamine secretion from mast cells. A number of rats were anesthetized and then injected intravenously with 5 nmol/kg* of neurotensin, whereas others (control rats) received intravenous injections of saline solution to control for the possibility that the act of injection itself pro-

*nmol/kg = nanomoles per kilogram.

```
voked histamine release. Some quantity of blood
was withdrawn from a number of rats 1, 3, 15, or
30 minutes after injections were made, and the
histamine concentration in each blood sample was
somehow determined.
```

Now, in a few short sentences, let us try to summarize the results beginning with the most general statements we can make.

1. Neurotensin had a dramatic and rapid effect on histamine concentrations in the blood of laboratory rats, with histamine concentration increasing by about 20 times within one minute after injection.
2. The effect seems rather short-lived; only two minutes into the study the histamine concentrations had already fallen considerably from the peak level recorded one minute earlier.
3. By 30 minutes after the injection, histamine concentration approached initial levels.
4. The effect shown is clearly due to the peptide rather than the procedure itself, because histamine concentrations in the blood of control rats receiving saline injections did not change appreciably during the study.

You can approach tabular data in the same way, first reading the accompanying legends and column headings, and then asking yourself *how* the study was done, *why* it might have been undertaken, and what the key general results seem to be; if the tables were prepared with as much care as the figure we just looked at, you should be able to answer each of these questions. Apply the same procedure figure by figure, table by table, until you reach the last bit of data. Now it is safe to actually read the text of the paper. Perhaps you will find the author reaching conclusions different from your own. Perhaps you missed some crucial element in studying the data yourself, or you will learn something crucial in the text of the paper that was not made clear in the figure. In this case, you can happily leave your own opinion behind and embrace the author's. On the other hand, you may genuinely disagree with the conclusion reached by the paper's author (or authors), or you may object to the author's interpretation because of some concern you have about the methodology used. This is, in fact, how new questions often get asked in science, and how new studies get designed, each new step building on the work of others. Most of us will never have the final say in any particular research area, but we can each contribute a valuable next step, even if our individual interpretation of that step turns out to be mildly or even dramatically wrong.

You may be on your way to a productive scientific career, research project in hand. At the very least, you will be in an excellent position to *discuss* the paper, on its own or in relationship to other studies that you will go on to scrutinize as thoroughly. If you don't go through these steps in "reading" the data, you will be all too accepting of the author's interpretation. In consequence, you will have considerable difficulty avoiding the book report format in your writing, simply repeating what others did and what they say they found. You can move your work to a higher, more interesting plane (more interesting for you and the reader) by becoming a brain-on reader. It's tough slogging, and although it becomes considerably easier with practice, it never becomes trivial. But it does become fun, and satisfying, in the same way that a good tennis game can be fun and satisfying, even when played in hot, humid weather.

READING TEXT

Resist the temptation to copy your source's words verbatim. Instead, try to summarize chunks of material as you read along. In that way, you will be processing the information as you read, and you will be one major step closer to having something interesting to write about later.

To summarize effectively, you must first determine the most important points in the material you wish to summarize. Consider the following paragraph from Rachel Carson's landmark book *Silent Spring.** First published in 1962, this book marks for many people the start of the environmental movement.

> Water, soil, and the earth's green mantle of plants make up the world that supports the animal life of the earth. Although modern man seldom remembers the fact, he could not exist without the plants that harness the sun's energy and manufacture the basic foodstuffs he depends upon for life. Our attitude toward plants is a singularly narrow one. If we see any immediate utility in a plant we foster it. If for any reason we find its presence undesirable or merely a matter of indifference, we may condemn it to destruction forthwith. Besides the various plants that are poisonous to man or his livestock, or crowd out food plants, many are marked for destruction merely because, according to our narrow view, they happen to be in the wrong place at the wrong time. Many others are destroyed merely because they happen to be associates of the unwanted plants.

What are the key points in this paragraph? What information would upset Ms. Carson were we to leave it out?

- All animals, including people, depend on plants for food.
- People think little about the consequences of destroying plants that annoy them or don't seem to do anything useful.

Here is a possible one-sentence summary that incorporates both these points:

*I have not altered the original wording, with its anachronistic use of what is now generally considered to be sexist writing. If Rachel Carson was writing this today, she would most likely replace "modern man" with "people" and "he" with "us". This would actually strengthen the paragraph, by pointing the finger clearly at all of us.

```
People destroy any plant that happens to annoy them
or doesn't seem to do anything useful, forgetting
the extent to which all animal life ultimately de-
pends on photosynthesis.
```

This summary is (1) accurate, (2) complete—it incorporates all major points, (3) self-sufficient—it makes good sense even if the reader has never read the original text, and (4) in my own words. Get in the habit of writing such summaries as you read and as you take notes in lecture. It is a real challenge, but one that gets easier with practice. Persist: the eventual payoff is tremendous.

PLAGIARISM AND NOTE-TAKING

The paper or report you submit for evaluation must be original: it must be *your* work. Submitting anyone else's work under your name is plagiarism and can get you expelled from college. Presenting someone else's ideas as your own is also plagiarism. Consider the following two paragraphs:

```
     Smith (1991) suggests that this discrepancy in
feeding rates may reflect differences in light in-
tensities used in the two different experiments.
Jones (1994), however, found that light intensity
did not influence the feeding rates of these ani-
mals and suggested that the rate differences re-
flect differences in the density at which the ani-
mals were held during the two experiments.
```

```
     This discrepancy in feeding rates might reflect
differences in light intensities. Jones (1994),
however, found that light level did not influence
```

feeding rates. Perhaps the difference in rates reflects differences in the density at which the animals were held during the two experiments.

The first example is fine. In the second example, however, the writer takes credit for the ideas of Smith and Jones; the writer has plagiarized.

Plagiarism sometimes occurs unintentionally through faulty note-taking. Photocopying an article or book chapter does not constitute note-taking; neither does copying a passage by hand, occasionally substituting a synonym for a word used by the source's author. Take notes using your own words; you must get away from being awed by other people's words and move toward building confidence in your own thoughts and phrasings. Note-taking involves critical evaluation; as you read, you must decide either that particular facts or ideas are relevant to your topic or that they are irrelevant. As Sylvan Barnet says in *A Short Guide to Writing About Art* (1989. Little, Brown and Company, third edition, p. 155), "You are not doing stenography; rather, you are assimilating knowledge and you are thinking, and so for the most part your source should be digested rather than engorged whole." If an idea is relevant, you should jot down a summary using your own words. Try not to write complete sentences as you take notes; this will help you avoid unintentional plagiarism later and will encourage you to see through to the essence of a statement while note-taking.

Sometimes the authors' words seem so perfect that you cannot see how they might be revised to best advantage for your paper. In this case, you may wish to copy a phrase or a sentence or two verbatim, but be sure to enclose this material in quotation marks as you write, and clearly indicate the source and page number from which the quotation derives. If you modify the original wording slightly as you take notes, you should indicate this as well, perhaps by using modified quotation marks: " . . . ". If your notes on a particular passage are in your own words, you should also indicate this as you write. I precede such notes, reflecting my own ideas or my own choice of words, with the word *Me* and a colon. If you take notes in this manner, you will avoid the unintentional plagiarism that occurs when you later forget who is actually responsible for the wording of your notes or who is actually responsible for the origin of an idea.

If you find yourself copying verbatim or paraphrasing your source, be sure it is not simply because you do not understand what you are reading. Be honest with yourself. It is always best to summarize in your own words as you read along; at the very least you should think your way to some good questions about what you are reading and write those questions down. Sooner or later, serious intellectual engagement is required; there are no shortcuts available here, I'm afraid.

Here is an example of some notes taken using the suggested system of notation. These notes are based on a paper published in 1989 by Mark D. Bertness. Figure 2.2 shows an excerpt from the paper, and Figure 2.3 shows some notes based on that excerpt. As shown in Figure 2.3, the note-taker has clearly distinguished between his or her thoughts and the author's thoughts, and between what the author has done and what the student thinks could be done later or might have influenced the results (for example, "Me: Why is it impt. to do study in protected bay?"). Note that the student has avoided using complete sentences, focusing instead on getting the basic points and pinning down a few words and phrases that might be useful later. This student will not have to worry about accidental plagiarism when writing a paper based on these notes. Moreover, the student is well on the way to preparing a solid essay since the style of note-taking indicates clearly that the student has been thinking while reading.

Some people suggest taking notes on index cards, with one idea per card so that the notes can be sorted readily into categories at a later stage of the paper's development. If you prefer to take notes on full-size paper, begin a separate page for each new source and write on only one side of each page to facilitate sorting later.

INTRODUCTION

The roles that intra- and interspecific competition play in shaping natural populations and communities has long been of interest to ecologists (Hairston et al. 1960, Strong et al. 1984). High population densities often result in competitive processes dictating natural distribution and abundance patterns (e.g., Buss 1986), but consumer pressure (Paine 1966, Harper 1969) and physical disturbance (Dayton 1971, Platt 1975) often reduce densities and minimize the importance of competition in nature. Limited recruitment (Underwood and Denley 1984, Roughgarden et al. 1987) and harsh physical conditions (Connell 1961*a*, Fowler 1986) also potentially limit population size, and reduce the impact of competitive phenomena on populations. High population densities, however, can also facilitate survival by buffering individuals from interspecific competitive pressures (Buss 1981), consumers (Atsatt and O'Dowd 1976), physical disturbance (Bertness and Grosholz 1985), and physiological stress (Hay 1981). Largely due to these contrasting effects of recruitment variation in natural populations and the variety of other factors that can independently or interactively influence populations, general statements relating recruitment density to population processes have been hard to make even in extensively studied systems (Connell 1985).

The study of plant and animal assemblages in hard substratum marine habitats has been valuable in understanding the interplay of biotic and physical factors in generating pattern in natural communities (e.g., Connell 1961*b*, Paine 1966, Dayton 1971). Sessile organisms in these systems are often space-limited, found along gradients of environmental harshness, and easily trackable and amenable to experimental manipulation. Acorn barnacles are conspicuous members of many temperate zone intertidal communities, dominating a distinct zone at intermediate to high tidal heights (Stephenson and Stephenson 1949). Within this habitat, barnacles compete intra- and interspecifically by crushing and/or overgrowing neighbors (Connell 1961*a*, Wethey 1983*a*), are limited from higher intertidal habitats by heat and desiccation (Connell 1961*a*, Wethey 1983*b*), and from lower intertidal habitats by predators and competitors (Connell 1972, Menge 1976).

In this paper, I examine the intraspecific density-dependent dynamics of the barnacle *Semibalanus balanoides* in a wave-sheltered southern New England bay. I document variation in recruitment and survivorship, investigate the role of density-dependent mortality in generating survivorship patterns across the intertidal habitat, and examine physical constraints on barnacle success.

METHODS

The *S. balanoides* population at the study site was examined in 10 × 10 cm quadrats on boulder (~1 m diameter) surfaces. A third of these quadrats were placed at low (−0.1 to 0.0 m, relative to mean low water) tidal heights, a third at intermediate (0.0 to +0.5 m) tidal heights, and a third at high (+0.5 to 1.0 m) tidal heights. Quadrats were individually numbered with aluminum tags attached to 6-mm (quarter-inch) stainless steel corner screws. Quadrats were established in February 1985 and followed until September 1987. All quadrats were photographed monthly from March–September through 1986. In 1987 photographs were taken in April, after settlement, and in September. The only major barnacle predator at the site, *Urosalpinx cinerea*, was removed from boulders containing quadrats in 1985 and 1986.

(margin markers: ① ② ③ ④ ⑤ ⑥ ⑦ ⑧)

Figure 2.2
Excerpt from M. D. Bertness. 1989. Intraspecific competition and facilitation in a northern acorn barnacle population. *Ecology* 70: 257–268. Numbers in the margin correspond to the notes shown in the margin of Figure 2.3.

1) Low no. individuals/m²* -----► reduced competition for resources.

Factors acting to keep pop. density low: physical stress, predation, low recruitment.

ME: What is "recruitment"? How defined?

2) Under some circumstances, survival is better when pop. densities are high; e.g., crowding can protect from physical stress. ME: Really? How?

3) Key issue: How do physical and biological factors interact to determine plant and animal distributions?

4) Acorn barnacles good to study since don't move around, are abundant intertidally, and are subject to various degrees of heat stress, dehydration stress, predation, and competition (including with other barnacles of same or other species) ME: How many barnacle species are there? How many on one beach?

5) This study deals with one species (Semibalanus balanoides) from New England coast (protected area, no waves).

ME: Why is it impt. to do study in protected bay? Would expect diff. results in more waveswept area? Why? Or just easier to work w/o** having to deal with waves?

Looks at recruitment (ME: appearance of young barnacles on rocks?)

[yes: see definit. at bottom left p. 258]

and survival at diff. tidal heights. Hopes to explain findings with respect to temp. stress.

6) Methods: Sets up 10 x 10 cm quadrats on boulders at 3 diff. tidal heights. ME: if in highest tide area, barnacles exposed longer to air.

7) Follows barnacles in quadrats for about 2.5 yrs, by monthly photos.

8) Removes the major predator (Urosalpinx cinerea) monthly in first two years. ME: U. cinerea = "oyster drill" snail.

Figure 2.3
Handwritten notes based on article by M. D. Bertness (see Figure 2.2). Numbers in the margin correspond to the indicated portions of Figure 2.2.

*m² = meters squared   **w/o = without

As you take notes, be sure to make a complete record of each source used: author(s), year of publication, volume and page numbers (if consulting a scientific journal), title of article or book, publisher, and total number of pages (if consulting a book). It is not always easy to relocate a source once it is returned to the library stacks; in fact, the source you forgot to record completely is always the one that vanishes as soon as you realize that you need it again. Furthermore, before you finish with a source, it is good practice to read the source through one last time to be sure that your notes accurately reflect the content of what you have read.

As another example of effective note-taking, consider some notes based on a paragraph from Charles Darwin's *The Origin of Species,* published in 1859. Accompanying the notes is the selection from Darwin's work on which the notes were based (Figure 2.4). The notes (Figure 2.5) were being taken for an essay on the mechanism of natural selection. Notice that the student has taken notes selectively, that the notes are generally not taken in complete sentences, and that the student has found it unnecessary to quote any of the material directly and has clearly distinguished his or her own thoughts from those of Darwin.

You cannot take notes in your own words if you do not understand what you are reading. Similarly, it is difficult to be selective in your note-taking until you have achieved a general understanding of the material. I suggest that you first consult at least one general reference text and read the material carefully, as recommended earlier. Once you have located a particularly promising scientific article, read the entire paper at least once without taking any notes. Resist the (strong) temptation to annotate and take notes during this first reading even though you may feel that without a pen in your hand you are accomplishing nothing. Put your pencils, pens, notecards, paper, or laptop computer away, and read. Read slowly and with care. Read to understand. Study the illustrations, figure captions, tables, and graphs carefully, and try to develop your own interpretations before reading those of the author(s). Don't be frustrated by not understanding the paper

It may be worth while to give another and more complex illustration of the action of natural selection. Certain plants ←① excrete sweet juice, apparently for the sake of eliminating something injurious from the sap: this is effected, for instance, by glands at the base of the stipules in some Leguminosæ, and at the backs of the leaves of the common laurel. This juice, though small in quantity, is greedily sought by insects; but their visits do not in any way benefit the plant. Now, let us ←② suppose that the juice or nectar was excreted from the inside of the flowers of a certain number of plants of any species. Insects in seeking the nectar would get dusted with pollen, and would often transport it from one flower to another. The flowers of two distinct individuals of the same species would thus get crossed; and the act of crossing, as can be fully proved, gives rise to vigorous seedlings which consequently would have the best chance of flourishing and surviving. The ←③ plants which produced flowers with the largest glands or nectaries, excreting most nectar, would oftenest be visited by insects, and would oftenest be crossed; and so in the long-run would gain the upper hand and form a local variety. The flowers, also, which had their stamens and pistils placed, in relation to the size and habits of the particular insects which visited them, so as to favour in any degree the transportal of the pollen, would likewise be favoured. We might have taken ←④ the case of insects visiting flowers for the sake of collecting pollen instead of nectar; and as pollen is formed for the sole purpose of fertilisation, its destruction appears to be a simple loss to the plant; yet if a little pollen were carried, at first occasionally and then habitually, by the pollen-devouring insects from flower to flower, and a cross thus effected, although nine-tenths of the pollen were destroyed it might still be a great gain to the plant to be thus robbed; and the individuals which produced more and more pollen, and had larger anthers, would be selected.

Figure 2.4
Taken from Charles Darwin's *The Origin of Species,* published in 1859, from which notes on the mechanism of natural selection were taken (see Figure 2.5).

at the first reading; understanding scientific literature takes time and patience—and often many rereadings, even for practicing biologists. Concentrate not only on the results reported in the papers but also on the reason the study was undertaken and the way the data were obtained. The results of a study are real; the inter-

Plant evolution tied to insect behavior. Flowers now = effective in attracting insects for pollen exchange; how originate?

① Some plant sap apparently toxic (me: no evidence given). Plant nectar makes sap less nasty. Insects attracted to the sweet nectar, even though produced by plant originally to protect plant.

② If plant produces nectar in flower, insects attracted to flower, thus transport pollen, facilitate cross-fert. me: note that selection for nectar prod. in flower can occur only if a few plants accidentally start secreting nectar in flowers. Note that nectar not orig. prod. to attract insects; selected to protect plant, but once being produced, can evolve different function.

③ Those flowers that have greatest success attracting insects will spread the most pollen. me: now know that the genes of these flowers would thus increase prob. of successful representation in next generation.

④ Nectar prod. not essential to explain evol. of insect-mediated cross pollination. Suppose insect feeds on pollen (as some spp. do). Some pollen would stick to legs and be transferred to another flower. Again, flowers most successful in attracting insects would incr. chance of spreading genes, even though most of the pollen gets eaten.

Figure 2.5
Handwritten notes based on the passage shown in Figure 2.4.

pretation of the results is always open to question. And the interpretation is largely influenced by the way the study was conducted. Read with a critical, questioning eye. Many of the interpretations and conclusions in today's journals will be modified in the future. It is difficult, if not impossible, to have the last word in Biology; progress is made by continually building on and modifying the work of others.

By the time you have completed your first reading of the paper, you may find that the article is not really relevant to your topic after all or is of little help in developing your theme. If so, the preliminary read-through will have saved you from wasted note-taking. If you have photocopied the article, all has not been lost; simply turn the pages over and use them for notepaper.

LOCATING USEFUL SOURCES

Once you have carefully read the appropriate sections of your textbook and the relevant portions of your class notes to get a solid foundation, you are ready to delve into more specialized material. When seeking relevant books, you may have to become a little devious before you can convince the library card catalog or computer search system to satisfy your request for information. Suppose, for example, that you wish to find material on reptilian respiratory mechanisms. You might try, to no avail, looking under Respiration or Reptiles, but looking under Physiology or Comparative Physiology will probably pay off. Similarly, in researching the topic of annelid locomotion, you might try, unprofitably, searching under Annelids, Locomotion, or Worms; looking under Invertebrate Zoology will probably turn up something useful. Using a library card file or computer search system is like using the Yellow Pages of the telephone book; the phone number of the local movie house isn't found under a heading of Movies or Movie Theaters, but under Theaters, not the first place I'd look. If at first you don't succeed. . . .

The references given in textbooks often provide good access to the primary research literature, as do those given in review articles. Five particular journals may contain especially useful reviews in this regard: *American Zoologist, Biological Reviews, Bio-*

Science, Scientific American, and *Quarterly Review of Biology.* It is also profitable to browse through recent issues of journals relevant to your topic; ask your instructor to name a few journals worth looking over. If you find an appropriate article in the recent literature, consult the literature citations at the end of the article for additional references worth consulting. This is an especially easy and efficient way to accumulate research material; the yield of good references is usually high for the amount of time invested.

Using *Science Citation Index*

Another way to track down pertinent recent references is by using the *Science Citation Index,* published by the Institute for Scientific Information (ISI) in Philadelphia. Because it is expensive, not all libraries subscribe to this service, so be sure to check with your reference librarian before getting your hopes up. Your reference librarian should also be able to assist you in using any of the services discussed in this chapter. For each of these services, brief but helpful user's guides appear at the beginning of each volume.

To use the *Science Citation Index,* you must first have discovered at least one paper (the so-called key paper) from the primary literature pertinent to your quest. Consulting the Citation volumes of the index, you look under the name of the author who wrote this key paper. Below that author's name, you will find a listing of additional references, one of which should be for the paper you have already read and found to be particularly appropriate to your topic. Beneath the listing for this reference, you will find a list of all the papers that have cited your key paper during the year covered by the index volume consulted. Suppose, for example, you have obtained and read the following reference, cited at the end of a chapter in your class textbook, and have determined the paper to be of special use in developing your topic:

Kandel, E. R. and J. H. Schwartz. 1982. Molecular biology of learning: Modulation of neurotransmitter release. *Science* 218: 433–443.

Looking in the Citation volume for 1991, you will find the listing shown in Figure 2.6, which includes nearly 20 of E. R. Kandel's papers, published between 1961 and 1986, each of which has been cited at least once in the most recent year or so. I have indicated the 1982 paper of interest to us by an arrow at the left of the reference. The bracket indicates six recent papers, listed in alphabetical order by author, that cite the Kandel and Schwartz reference. The list includes, for example, the following item:

ICHINOSE, M. BRAIN RES 549 146 91.

This tells us that a paper citing the Kandel and Schwartz (1982) article was published in 1991 by M. Ichinose in volume 549 of the journal *Brain Research,* beginning on page 146. Most likely, a paper that cites your key reference in its bibliography is appropriate to your topic and is therefore worth consulting. If you wished to have more information before going to find volume 549 of *Brain Research,* you could look up the complete citation of the paper, including its full title, in the Source Index volume of *Science Citation Index.*

Looking elsewhere in Figure 2.6, note that most journals are not listed by their full names. To avoid confusion, each volume of the *Science Citation Index* begins with an explanation of all journal abbreviations used. You will find, for example, that IMM CELL B stands for a journal called *Immunology and Cell Biology,* whereas CAN J MICRO stands for the *Canadian Journal of Microbiology.*

Science Citation Index is published at two-month intervals and is consolidated into a smaller number of volumes yearly.

Using *Zoological Record*

Another useful indexing service is the *Zoological Record,* published jointly by BIOSIS and the Zoological Society of London.

KANDA M VOL PG YR
88 JPN J GENET 63 127
CHEN LFO BOTAN B A S 32 153 91
89 ANN PROBAB 17 379
KANDA M NAG MATH J 122 63 91 N
89 J BIOCHEM-TOKYO 105 653
SCHOLTEN JD SCIENCE 253 182 91
89 SINGAPORE PROB C
KANDA M NAG MATH J 122 63 91 N
KANDA N
83 P NATL ACAD SCI USA 80 4069
HASUMI K BIOC BIOP A 1083 289 91
87 CANCER RES 47 3291
HIRANO Y FEBS LETTER 284 235 91
KANDA P
80 J LIPID RES 21 257
SONNET PE CHEM PHYS L 58 35 91
KANDA S
67 J POLYM SCI C 17 151
ASWAR AS COLLOID P S 269 547 91
81 J PHYSIOL-LONDON 299 127
STERIN ABF ARCH I PHYS 99 141 91
81 PHYS CHEM BIOL 25 167
AZUMA Y CLIN CHEM 37 1132 91 L
KANDA T
** UNPUB J PROPULSION P
KANDA T J PROPUL P 7 431 91
79 MOSQ NEWS 39 568
FOLEY DH J AUS ENT S 30 109 91
83 ACTA OTOLARYNGOL S S 393 25
YOSHIHAR T ACT OTO-LAR 111 607 91
86 BIOCHEMISTRY-US 25 1159
KITAHATA S BIOCHEM 30 6769 91
86 J GEN APPL MICROBIOL 32 541
BINNINGE DM MOL G GENE 227 245 91
87 EAR RES JPN 18 501
KANDA T ACT OTO-LAR 97 91
87 J VIROL 62 610
87 JPN J CANCER RES 78 103
PHELPS WC CURR T MICR 144 153 89 R
88 J VIROL 62 610
FIRZLAFF JM P NAS US 88 5187 91
HASHIDA T J GEN VIROL 72 1569 91
IFTNER T CURR T MICR 144 167 89 R
KLEINHEI.A " 144 175 89 R
WILBUR DC YALE J BIOL 64 113 91
88 VIROLOGY 165 321
MUNGER K J VIROLOGY 65 3943 91 N
SCHEFFNE.M P NAS US 88 5523 91
TAKAMI Y INT J CANC 48 516 91
89 CLIN NEUROPATHOL 8 34
FUKUTANI Y J NEUROL 238 191 91
89 CLIN NEUROPATHOL 8 134
KANDA T J NEUR SCI 103 42 91
89 MOL GEN GENET 216 526
HASEBE K MYCOLOGIA 83 354 91
89 NAL TR1002 NAT AER L
KANDA T J PROPUL P 7 431 91
91 IN PRESS BRAIN 114
KANDA T J NEUR SCI 103 42 91
KANDA Y
74 J BIOL CHEM 249 6796
HARMS PJ COMP BIOC B 99 239 91
82 IEEE T ELECTRON DEV 29 64
DONZIER E SENS ACTU-A 26 357 91
82 IEEE T ELECTRON DEV 29 151
MUNTER PJA SENS ACTU-A 27 747 91
86 METALLOGRAPHY 19 461
TAKAO Y J NUCL MAT 179 298 91
87 JPN J APPL PHYS PT 1 26 1031
CHUNG GS ELECTR LETT 27 1098 91
88 ANAL CHIM ACTA 207 269
VANVEEN EH ANALYT CHE 63 1441 91
90 ANAL CHEM 62 2084
FOX DL ANALYT CHE 63 R292 91
KANDA Z
76 CHEM LETT 199
78 INORG CHEM 17 910
OCONNOR JM J AM CHEM S 113 4530 91
KANDALA CVK
87 T ASAE 30 793
88 INT AGROPHYSICS 4 3
88 T ASAE 31 1890
NELSON SO T ASAE 34 507 91
89 J AGR ENG RES 44 125
NELSON SO T ASAE 34 507 91
" " 34 513 91
KANDALAF N
82 JAMA-J AM MED ASSOC 248 2166
SOTANIEM KA J NE NE PSY 54 645 91 N
KANDALL S
77 EARLY HUM DEV 1 1
WITTMANN BK INT J ADDIC 26 213 91
KANDALL SR
75 ADDICT DIS INT 2 347
WITTMANN BK INT J ADDIC 26 213 91
77 EARLY HUM DEV 1 159
DOBERCZA.TM J PEDIAT 118 933 91
83 AM J DIS CHILD 137 378
DOBERCZA.TM J PEDIAT 118 933 91
HOEGERMA.G CLIN PERIN 18 51 91
KANDARAKIS ED
85 NEUROPEPTIDES 8 21
JARVINEN A PHARM TOX 68 371 91
KANDASAMY K
** IN PRESS Z PHYS CHEM
SAKAMOTO Y SCR MET MA 25 1629 91
88 Z PHYS CHEM 158 253
89 Z PHYS CHEM NEUE FOL 163 41

KANDEL DB VOL PG YR
86 ARCH GEN PSYCHIAT 43 255
MOREAU D J AM A CHIL 30 642 91
86 ARCH GEN PSYCHIAT 43 746
WALTER HJ J AM A CHIL 30 556 91
87 INT J ADDICT 22 319
GREEN G BR J ADDICT 86 745 91
KANDEL E
78 VOPR NEIROKHIR 1 51
SALEH J NEUROSURGE 29 113 91
91 IN PRESS COLD SPRING 55
EDELMAN GM ANN R BIOCH 60 155 91 R
KANDEL EI
77 J NEUROSURG 46 12
SISTI MB J NEUROSUR 75 40 91
→ KANDEL ER
61 J NEUROPHYSIOL 24 225
DOZE VA NEURON 6 889 91
ISOKAWA M BRAIN RES 551 94 91
61 J NEUROPHYSIOL 24 243
ALBOWITZ B EUR J NEURO 3 570 91
DAVENPOR.R BIOL CYBERN 65 47 91
LACROIX D J PHARM EXP 257 1081 91
66 J PHYSIOL-LONDON 183 287
FERGUSON GP J EXP BIOL 158 63 91
68 ELECTROPHYSIOLOGICAL 385
72 RES PUBLICATIONS ASS 50 91
KUPFERMA.I PHYSIOL REV 71 683 91 R
76 CELLULAR BASIS BEHAV
ARSHAVSK.YI DAN SSSR 318 232 91
COLEBROO.E CELL MOL N 11 305 91
FERGUSON GP J EXP BIOL 158 63 91
" " 158 97 91
VENABLE N BEHAV PHAR 2 161 91
79 BEHAVIORAL BIOL APLY
FERGUSON GP J EXP BIOL 158 63 91
" " 158 97 91
79 HARVEY LECT 73 19
ISEKI M JPN J A P 1 30 1117 91
79 HARVEY LECTURES SERI 73 19
SAKHAROV DA J EVOL BIOC 26 557 90 D
80 CELLULAR BASIS BEHAV
ARSHAVSK.YI NEUROPHYSIO 22 579 90
" " 22 587 90
KUDRASHO.IV " 22 525 90
81 NATURE 293 697
RICHES IP J NEUROSC 11 1763 91
81 PRINCIPLES NEURAL SC
FRIEDMAN R J AM GER SO 39 650 91
81 PRINCIPLES NEURAL SC 731
STELLA A PHYSIOL REV 71 659 91 R
→ 82 SCIENCE 218 433
ASZODI A P NAS US 88 5832 91
BUCHNER E J NEUROGEN 7 153 91 R
FABER DS BRAIN BEHAV 37 288 91
ICHINOSE M BRAIN RES 549 146 91 N
MASSICOT.G NEUROSCI B 15 415 91 R
SCHWARTZ JH BIOCH SOC T 19 387 91
83 AM J PSYCHIAT 140 1277
HARRISON PJ J NERV MENT 179 309 91
85 PRINCIPLES NEURAL SC
JULESZ B REV M PHYS 63 735 91 R
MUSILA M BIOSYSTEMS 25 179 91
85 PRINCIPLES NEURAL SC 132
TAKAHASH.M BRAIN RES 551 279 91
86 FIDIA RES F NEUROSCI 7
BRAVAREN.NI ZH VYSS NER 41 107 91
86 NEUROSCIENCE AWARD L 7
ARSHAVSK.YI DAN SSSR 318 232 91
86 NEUROSCI RES 3 498
GARCIAGI.M COMP BIOC B 99 65 91
MASSICOT.G NEUROSCI B 15 415 91 R
87 SYNAPTIC FUNCTION 471
BUCHNER E J NEUROGEN 7 153 91 R
89 J NEUROPSYCHIAT 1 103
HARRISON PJ J NERV MENT 179 309 91
89 PRINCIPLES NEUROSCIE 620
VRENSEN GFJ EXP EYE RES 52 647 91
KANDEL G
83 ARCH INTERN MED 143 1400
JAIN A CHEST 99 1403 91
SUMIMOTO T AM HEART J 122 27 91
KANDEL J
74 J BIOL CHEM 249 2088
LOBBAN MD BIOC BIOP A 1078 155 91
SEABROOK RN EUR J BIOCH 198 741 91
KANDEL M
78 J BIOL CHEM 252 679
SILVERMA.DN CAN J BOTAN 69 1070 91
78 J BIOL CHEM 253 679
HUSIC HD CAN J BOTAN 69 1079 91
KANDEL RA
87 BIOCHEM INT 15 1021
90 BIOCHIM BIOPHYS ACTA 1053 130
CRUZ TF BIOCHEM J 277 327 91
90 J RHEUMATOL 17 953
BALBLANC JC REV RHUM 58 343 91
CRUZ TF BIOCHEM J 277 327 91
DEAN DD SEM ARTH R 20 2 91
KANDEL SN
73 CLIN ORTHOP RELAT R 96 108
SMYTH GB CORNELL VET 81 239 91
KANDELAKI HS
87 BIOCHIM BIOPHYS ACTA 170 893
AGOSTIAN.A J PHOTOCH A 58 201 91
KANDELAKI TS
73 MIKROELEKTRONIKA 2 259
VLASOV SI IVUZ FIZ 34 121 91 N
KANDELER E

KANDIL A VO
SOLIMAN FM PHOSPHOR S 6
KANDIL AT
75 J INORG NUCL CHEM 37 229
DUKOV IL ACT CHIM HU 12
SAAD EA MICROCHEM J 4
80 J INORG NUCL CHEM 42 149
DUKOV IL ACT CHIM HU 12
KANDIL E
70 BRIT J DERMATOL 83 405
FETSCH JF MOD PATHOL
KANDIL OM
87 FED PROC 46 441
AQEL MB J ETHNOPHAR 3
KANDIL SH
86 J MATER SCI LETT 5 112
AFIFY N J NON-CRYST 12
KASSEM ME J THERM ANA 3
KANDILE NG
** IN PRESS ACTA CHIM H
** UNPUB ACTA CHIM HUNG
ISMAIL MF ACT CHIM HU 12
KANDILOV NK
63 IZVESTIYA AKADEM BMN 6
LANDSBER.JH SYST PARAS 1
KANDIYOTI R
84 FUEL 63 1583
IBARRA JV FUEL 7
KANDLER O
01 BERGEYS MANUAL SYSTE 2 1
TOBA T LETT APPL M 1
68 J BACTERIOL 96 1935
SIJTSMA L NETH MILK D
80 BIOCH PLANTS 3 221
STAHL B ANALYT CHE
82 ENCY PLANT PHYSL A 13 348
CRESPI MD PLANT PHYSL
83 A VAN LEEUW J MICROB 49 2
84 BERGEYS MANUAL SYSTE 2 1
TSENG CP J BACT 17
84 NATURWISSENSCHAFTLIC 37 9
SEAWARD MRD LICHENOLOGI 2
85 BACTERIA TREATISE ST 7 41
JOHANNSE.L INFEC IMMUN
85 KORTZFLEISCH 20
KANDLER O WATER A S P
86 ARCHAEBACTERIA 85 198
KOCH AL FEMS MIC R
86 BERGEYS MANUAL SYSTE 1
BUNCIC S J FOOD PROT 5
86 BERGEYS MANUAL SYSTE 2 1
JONSSON E J APPL BACT 7
KENNES C APPL MICR B 3
MOLLET B J BACT 17
86 BERGEYS MANUAL SYSTE 2 12
NGUYENTH.C J APPL BACT 7
PILONE GJ AM J ENOL V 4
POSNO M APPL ENVIR
87 AFZ 42 715
EVERS FH WATER A S P
KANDLER O "
MONCHAUX P "
88 ALLG FORST JAGDZTG 159 179
KANDLER O WATER A S P 5
89 12 INT S MOL BEAMS
KANDLER O Z PHYS D 1
89 12 INT S MOL BGEAMS
LEISNER T Z PHYS D 20
89 1 P STAT SEM PBWU FO 7
KANDLER O WATER A S P 54
90 THESIS U KONSTANZ
LEISNER T Z PHYS D 2
KANDLER P
1862 CODICE DIPLOMATICO I
DEGASPER.C TECTONOPHY 19
86 BERGEYS MANUAL SYSTE 1
LEBRAS G EUR J BIOCH 1
KANDLER R
26 WISS MEERESUNTERS KO 16
GUSTAFSO.RG AQUACULTUR 9
90 LANCET 335 469
DHUNA A NEUROLOGY 4
KANDOLF R
85 J MOL CELL CARDIOL 17 167
85 P NATL ACAD SCI USA 82 4818
REIMANN BY J VIROLOGY 6
87 CIRCULATION 76 262
MATOBA Y JPN CIRC J 5
88 COXSACKIEVIRUSES GEN 29
89 CONCEPTS VIRAL PATHO 3 28
89 SPRINGER SEMIN IMMUN 11
REIMANN BY J VIROLOGY 65
90 NEW ASPECTS POSITIVE 340
ROSE NR IMMUNOL TO 12
KANDOLO K
85 J CLIN MICROBIOL 21 980
MCGAREY DJ EXPERIENTIA 4
KANDOOLE BF
** ADMARCS COST STRUCTU
SAHN DE FOOD POLICY 1
KANDORI H
** UNPUB
KOBAYASH.T CHEM P LETT 18
89 BIOCHEMISTRY-US 28 6460
89 BIOPHYS J 56 453
89 PHOTOCHEM PHOTOBIOL 49 181
SHICHIDA Y BIOCHEM 30
KANDORI K

Figure 2.6
Detail of a page of the Citation volume from the 1991 *Science Citation Index.* The lower arrow indicates the key paper on which our search is based, a 1982 paper in *Science* (volume 218, beginning on page 433).

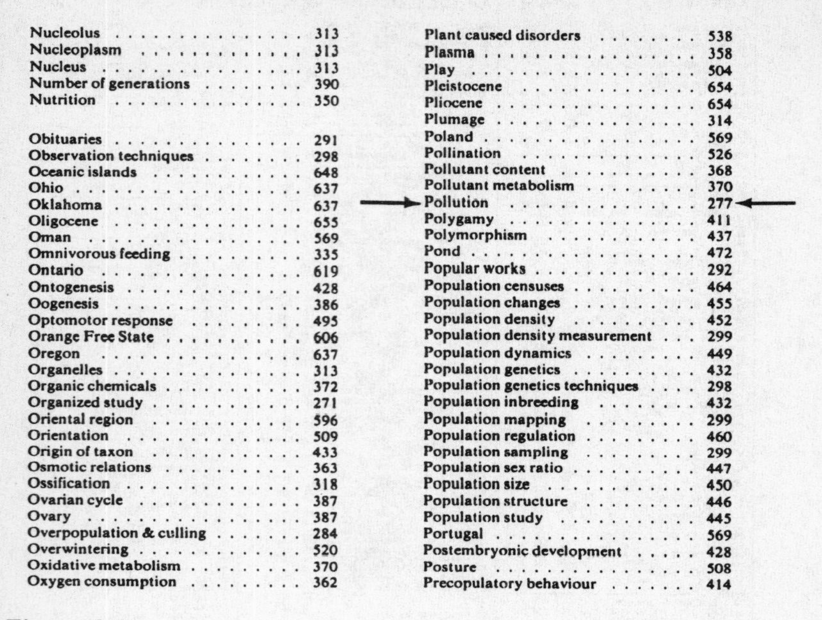

Subject	Page
Nucleolus	313
Nucleoplasm	313
Nucleus	313
Number of generations	390
Nutrition	350
Obituaries	291
Observation techniques	298
Oceanic islands	648
Ohio	637
Oklahoma	637
Oligocene	655
Oman	569
Omnivorous feeding	335
Ontario	619
Ontogenesis	428
Oogenesis	386
Optomotor response	495
Orange Free State	606
Oregon	637
Organelles	313
Organic chemicals	372
Organized study	271
Oriental region	596
Orientation	509
Origin of taxon	433
Osmotic relations	363
Ossification	318
Ovarian cycle	387
Ovary	387
Overpopulation & culling	284
Overwintering	520
Oxidative metabolism	370
Oxygen consumption	362
Plant caused disorders	538
Plasma	358
Play	504
Pleistocene	654
Pliocene	654
Plumage	314
Poland	569
Pollination	526
Pollutant content	368
Pollutant metabolism	370
Pollution	277
Polygamy	411
Polymorphism	437
Pond	472
Popular works	292
Population censuses	464
Population changes	455
Population density	452
Population density measurement	299
Population dynamics	449
Population genetics	432
Population genetics techniques	298
Population inbreeding	432
Population mapping	299
Population regulation	460
Population sampling	299
Population sex ratio	447
Population size	450
Population structure	446
Population study	445
Portugal	569
Postembryonic development	428
Posture	508
Precopulatory behaviour	414

Figure 2.7
Detail of a page of subject listings from the 1990 Chordata volume of the *Zoological Record.* The arrow indicates our topic of interest, directing us to page 277.

Zoological Record is published once a year in several volumes, each devoted to a particular animal phylum or group of related phyla. You will, for example, find separate volumes devoted to the Mollusca, the Annelida, and the Chordata. At the front of each volume is a section arranged by subject, as shown in the example in Figure 2.7 taken from the 1990 volume covering the members of our own phylum, the Chordata.

Suppose we wish to find some references concerning the influence of acid rain on the reproductive success of birds. As indicated by the arrows in Figure 2.7, the subject of Pollution is probably our best bet, so we turn to page 277 of the same volume. On page 277, we find a variety of interesting pollution-related references (see the circled heading in Figure 2.8), most of

Population changes relationships, Pennsylvania
GOODRICH, L.J., ET AL (3610)

POLLUTION

Abundance/biomass comparison indications, Netherlands (marine)
MEIRE, P.M., ET AL (6261)
Biological effects, England
Cygnus olor GLASER, G.A. (3515)
Breeding population decline relationship, lake, Northern Ireland
Melanitta nigra PARTRIDGE, J.K. (7096)
Compilation of entangled species in beach litter survey, West Germany
LIEDTKE, G., ET AL (5650)
Conservation implications, Turkey AKCAKAYA, H.R. (248)
Conservation role of RSPB, review, United Kingdom SAMSTAG, T. (8095)
Discarded fishing tackle, ropes & plastic litter
Mortality due to entrapment, seabirds, Netherlands
CAMPHUYSEN, C.J. (1656)
Habitat damage indicator & conservational significance, West Germany
Acrocephalus scirpaceus ANON (6)
Habitat utilization relationships, lake, Japan
Anatinae SUGIMORI, F., ET AL (8935)
Indicator, restoration aims, Great Lakes
Larus argentatus GILBERTSON, M. (3485)
Lake eutrophication, endangered status, West Germany
Tachybaptus ruficollis REICHHOLF, J. (7713)
Marine habitat, indicator role BATTY, L. (712)
Plastic debris incorporation into nests, Maine
Phalacrocorax auritus PODOLSKY, R.H., ET AL (7408)
Review, North America LEIGHTON, F.A. (5554)
Wetland population changes, indicator value, Michigan
MAINONE, R., ET AL (5905)

CHEMICAL POLLUTION

Accidental poisoning with fenthion, Orange Free State
Buteo buteo COLAHAN, B.D., ET AL (2023)
Acid rain effect on forest & breeding population changes, W Germany
OELKE, H. (6896)
Acid rain, effects on aquatic species, Europe & North America
DIAMOND, A.W. (2545)
→ Acid rain, relations with egg shell inferior quality, forest, Netherlands
Parus major DRENT, P.J., ET AL (2690)
Acute lead poisoning effect on tissue lead levels & blood values
Cygnus olor O'HALLORAN, J., ET AL (6877)
Agrichemicals in prairie wetlands potential effects & management, Canada
Anatidae FORSYTH, D.J. (3200)
Agrichemicals in prairie wetlands potential effects & management, USA
Anatidae GRUE, C.E., ET AL (3808)
Biological effects, past, present & future, overview
Falco peregrinus RATCLIFFE, D.A. (7652)
Blood protoporphyrin levels use as indicator of lead poisoning
Callonetta leucophrys PASSER, E.L., ET AL (7105)
Breeding population change relationships, Switzerland
Falco tinnunculus KAESER, G., ET AL (4929)
Tyto alba KAESER, G., ET AL (4929)

Figure 2.8
Detail of page 277 from *Zoological Record,* showing references on chemical pollution.

which were published in either 1989 or 1990, the timespan covered by this particular volume of *Zoological Record.* Under the heading "Chemical Pollution," a paper by Drent *et al.* (see arrow) seems particularly promising, and we are referred to reference no. 2690. The complete reference for this paper, including the names of all co-authors, the title of the paper, and the complete page numbers on which the article appears, is given in a separate section toward the front of the volume, as shown in Figure 2.9. Note the arrow in front of the citation for the Drent *et al.* article. These references are listed alphabetically and numerically so that we could have looked up the reference under Drent rather than by number.

In addition to the author and subject indexes already discussed, each volume of *Zoological Record* contains a Geographical Index (in which references are arranged by geographical area and country), a Systematic Index (in which articles are arranged by taxonomic group) and a Paleontological Index covering animals known only as fossils (arranged by geological epoch and era).

Using *Biological Abstracts*

Another widely used service is *Biological Abstracts,* published by BIOSIS in Philadelphia. *Biological Abstracts* is the most up-to-date of the published indexing services, but it can be frustrating to use. You begin in a straightforward manner, by looking up key words relevant to the topic being researched. Suppose, for example, you are looking for papers discussing the molecular basis for the action of prolactin, the pituitary hormone that triggers milk production in mammals. Key words, such as prolactin, are listed in the central portion of each column, as shown in Figure 2.10.

Unfortunately, an appropriate key word doesn't always lead to a useful reference. To the right and left of the key word *prolactin,* for example, and for the 50 lines under this key word, you will find a few words or parts of words (drawn from each paper's title with terms added by BIOSIS editors) that partially inform us

Drake, O.F. (2686)
Continental starlings in Devon.
Devon Birds **41**(4) 1988: 73-75, illustr. [In English]

Dransfeld, H. *see* Bangjord, G.

Dranzoa, C. & Rodrigues, R. (2687)
Two new records for Uganda.
Scopus **14**(1) 1990: 32-33. [In English]

Draulans, D. (2688)
Timing of breeding and nesting success of raptors in a newly colonized area in north-east Belgium.
Gerfaut **78**(4) 1988: 415-420, illustr. [In English with Dutch & French summaries]

Draulans, D. & van Vessem, J. (2689)
Some aspects of population dynamics and habitat choice of grey herons (*Ardea cinerea*) in fish-pond areas.
Gerfaut 77(4) 1987: 389-404, illustr. [In English with Dutch & French summaries]

Dreissens, G. *see* van der Burg, E.

→**Drent, P.J. & Woldendorp, J.W.** (2690)
Acid rain and eggshells.
Nature (Lond) **339** No 6224 1989: 431, illustr. [In English]

Drent, R. & Klaassen, M. (2691)
Energetics of avian growth: the causal link with BMR and metabolic scope.
NATO Adv Study Inst Ser Ser A Life Sci **173** 1989: 349-359, illustr. [In English]

Drent, R.H. *see* de Boer, W.F.
Drewes, L.A. *see* Flammer, K.

Drgonova, N. & Janiga, M. (2692)
Nest structure of alpine accentors (*Prunella collaris*) (Scop., 1769) in the low Tatras.
Biologia (Bratisl) **44**(10) 1989: 983-993, illustr.
[In English with Russian & Slovak summaries]

Drickamer, L.C. (2693)
Pheromones: behavioral and biochemical aspects.
Adv Comp Environ Physiol **3** 1989: 269-348, illustr. [In English]

Driel, F. van *see* van den Bergh, L.M.J.
Driesch, A. von den *see* Boessneck, J.
Driessens, G. *see* Buys, P.; van der Burg, E.

Drillon, V. (2694)
Analyse des causes de regression du grand tetras dans le massif de la Haute Meurthe.
Ciconia **13**(1-2) 1989: 11-18, illustr. [In French with English summary]

Drimmelen, B. van *see* Munro, W.T.

Figure 2.9
Detail of a page from the author index at the front of *Zoological Record,* with an arrow indicating the complete reference for the paper by Drent and Woldendorp.

of each paper's content. Each of these papers has something to do with prolactin, and the trick is to figure out what.

The only way to tell whether or not a reference is relevant to our topic is to look it up. Such listings provide many ambiguous

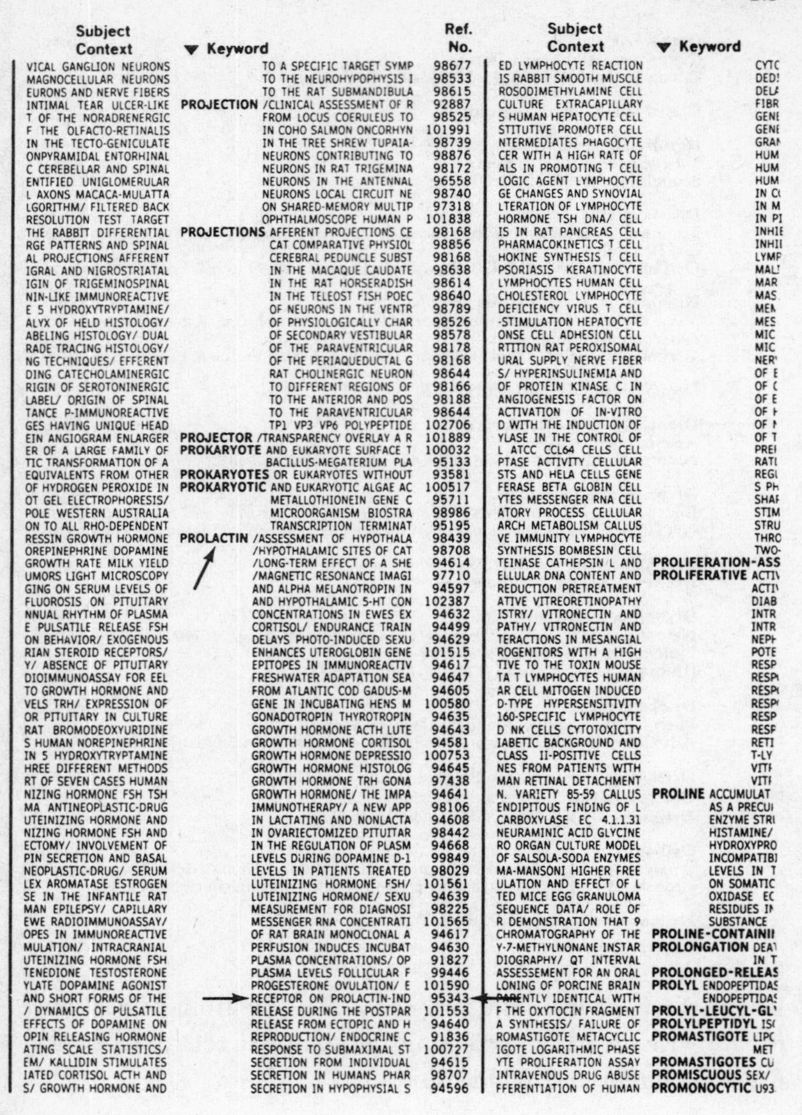

Bio

Subject Context	▼ Keyword		Ref. No.	Subject Context	▼ Keyword	
VICAL GANGLION NEURONS		TO A SPECIFIC TARGET SYMP	98677	ED LYMPHOCYTE REACTION		CYTC
MAGNOCELLULAR NEURONS		TO THE NEUROHYPOPHYSIS I	98533	IS RABBIT SMOOTH MUSCLE		DED!
EURONS AND NERVE FIBERS		TO THE RAT SUBMANDIBULA	98615	ROSODIMETHYLAMINE CELL		DELA
INTIMAL TEAR ULCER-LIKE	**PROJECTION**	/CLINICAL ASSESSMENT OF R	92887	CULTURE EXTRACAPILLARY		FIBR
T OF THE NORADRENERGIC		FROM LOCUS COERULEUS TO	98525	S HUMAN HEPATOCYTE CELL		GENE
F THE OLFACTO-RETINALIS		IN COHO SALMON ONCORHYN	101991	STITUTIVE PROMOTER CELL		GENE
IN THE TECTO-GENICULATE		IN THE TREE SHREW TUPAIA-	98739	NTERMEDIATES PHAGOCYTE		GRAN
ONPYRAMIDAL ENTORHINAL		NEURONS CONTRIBUTING TO	98876	CER WITH A HIGH RATE OF		HUM
C CEREBELLAR AND SPINAL		NEURONS IN RAT TRIGEMINA	98172	ALS IN PROMOTING T CELL		HUM
ENTIFIED UNIGLOMERULAR		NEURONS IN THE ANTENNAL	96558	LOGIC AGENT LYMPHOCYTE		HUM
L AXONS MACACA-MULATTA		NEURONS LOCAL CIRCUIT NE	98740	GE CHANGES AND SYNOVIAL		IN C(
LGORITHM/ FILTERED BACK		ON SHARED-MEMORY MULTIP	97318	LTERATION OF LYMPHOCYTE		IN M
RESOLUTION TEST TARGET		OPHTHALMOSCOPE HUMAN P	101838	HORMONE TSH DNA/ CELL		IN PI
THE RABBIT DIFFERENTIAL	**PROJECTIONS**	AFFERENT PROJECTIONS CE	98168	IS IN RAT PANCREAS CELL		INHIE
RGE PATTERNS AND SPINAL		CAT COMPARATIVE PHYSIOL	98856	PHARMACOKINETICS T CELL		INHII
AL PROJECTIONS AFFERENT		CEREBRAL PEDUNCLE SUBST	98168	HOKINE SYNTHESIS T CELL		LYMP
IGRAL AND NIGROSTRIATAL		IN THE MACAQUE CAUDATE	98638	PSORIASIS KERATINOCYTE		MAL!
IGIN OF TRIGEMINOSPINAL		IN THE RAT HORSERADISH	98614	LYMPHOCYTES HUMAN CELL		MAR
NIN-LIKE IMMUNOREACTIVE		IN THE TELEOST FISH POEC	98640	CHOLESTEROL LYMPHOCYTE		MAS.
E 5 HYDROXYTRYPTAMINE/		OF NEURONS IN THE VENTR	98789	DEFICIENCY VIRUS T CELL		MEN
ALYX OF HELD HISTOLOGY/		OF PHYSIOLOGICALLY CHAR	98526	-STIMULATION HEPATOCYTE		MES
ABELING HISTOLOGY/ DUAL		OF SECONDARY VESTIBULAR	98578	ONSE CELL ADHESION CELL		MIC
RADE TRACING HISTOLOGY/		OF THE PARAVENTRICULAR	98178	RTITION RAT PEROXISOMAL		MIC
NG TECHNIQUES/ EFFERENT		OF THE PERIAQUEDUCTAL G	98168	URAL SUPPLY NERVE FIBER		NER'
DING CATECHOLAMINERGIC		RAT CHOLINERGIC NEURON	98644	S/ HYPERINSULINEMIA AND		OF E
RIGIN OF SEROTONINERGIC		TO DIFFERENT REGIONS OF	98166	OF PROTEIN KINASE C IN		OF (
LABEL/ ORIGIN OF SPINAL		TO THE ANTERIOR AND POS	98188	ANGIOGENESIS FACTOR ON		OF E
TANCE P-IMMUNOREACTIVE		TO THE PARAVENTRICULAR	98644	ACTIVATION OF IN-VITRO		OF H
GES HAVING UNIQUE HEAD		TP1 VP3 VP6 POLYPEPTIDE	102706	D WITH THE INDUCTION OF		OF N
EIN ANGIOGRAM ENLARGER	**PROJECTOR**	/TRANSPARENCY OVERLAY A R	101889	YLASE IN THE CONTROL OF		OF T
ER OF A LARGE FAMILY OF	**PROKARYOTE**	AND EUKARYOTE SURFACE T	100032	L ATCC CCL64 CELLS CELL		PREI
TIC TRANSFORMATION OF A		BACILLUS-MEGATERIUM PLA	95133	PTASE ACTIVITY CELLULAR		RATI
EQUIVALENTS FROM OTHER	**PROKARYOTES**	OR EUKARYOTES WITHOUT	93581	STS AND HELA CELLS GENE		REGI
OF HYDROGEN PEROXIDE IN	**PROKARYOTIC**	AND EUKARYOTIC ALGAE AC	100517	FERASE BETA GLOBIN CELL		S PH
OT GEL ELECTROPHORESIS/		METALLOTHIONEIN GENE C	95711	YTES MESSENGER RNA CELL		SHAF
POLE WESTERN AUSTRALIA		MICROORGANISM BIOSTRA	98986	ATORY PROCESS CELLULAR		STIM
ON TO ALL RHO-DEPENDENT		TRANSCRIPTION TERMINAT	95195	ARCH METABOLISM CALLUS		STRU
RESSIN GROWTH HORMONE	**PROLACTIN**	/ASSESSMENT OF HYPOTHALA	98439	VE IMMUNITY LYMPHOCYTE		THRO
OREPINEPHRINE DOPAMINE		/HYPOTHALAMIC SITES OF CAT	98708	SYNTHESIS BOMBESIN CELL		TWO-
GROWTH RATE MILK YIELD		/LONG-TERM EFFECT OF A SHE	94614	TEINASE CATHEPSIN L AND	**PROLIFERATION-ASS**	
UMORS LIGHT MICROSCOPY		/MAGNETIC RESONANCE IMAGI	97710	ELLULAR DNA CONTENT AND	**PROLIFERATIVE**	ACTIV
GING ON SERUM LEVELS OF		AND ALPHA MELANOTROPIN IN	94597	REDUCTION PRETREATMENT		ACTIV
FLUOROSIS ON PITUITARY		AND HYPOTHALAMIC 5-HT CON	102387	ATIVE VITREORETINOPATHY		DIAB
NNUAL RHYTHM OF PLASMA		CONCENTRATIONS IN EWES EX	94632	ISTRY/ VITRONECTIN AND		INTR
E PULSATILE RELEASE FSH		CORTISOL/ ENDURANCE TRAIN	94499	PATHY/ VITRONECTIN AND		INTR
ON BEHAVIOR/ EXOGENOUS		DELAYS PHOTO-INDUCED SEXU	94629	TERACTIONS IN MESANGIAL		NEPH
RIAN STEROID RECEPTORS/		ENHANCES UTEROGLOBIN GENE	101515	ROGENITORS WITH A HIGH		POTE
Y/ ABSENCE OF PITUITARY		EPITOPES IN IMMUNOREACTIV	94617	TIVE TO THE TOXIN MOUSE		RESP
DIOIMMUNOASSAY FOR EEL		FRESHWATER ADAPTATION SEA	94647	TA T LYMPHOCYTES HUMAN		RESP
TO GROWTH HORMONE AND		FROM ATLANTIC COD GADUS-M	94605	AR CELL MITOGEN INDUCED		RESP
VELS TRH/ EXPRESSION OF		GENE IN INCUBATING HENS M	100580	D-TYPE HYPERSENSITIVITY		RESP
OR PITUITARY IN CULTURE		GONADOTROPIN THYROTROPIN	94635	160-SPECIFIC LYMPHOCYTE		RESP
RAT BROMODEOXYURIDINE		GROWTH HORMONE ACTH LUTE	94643	D NK CELLS CYTOTOXICITY		RESP
S HUMAN NOREPINEPHRINE		GROWTH HORMONE CORTISOL	94581	IABETIC BACKGROUND AND		RETI
IN 5 HYDROXYTRYPTAMINE		GROWTH HORMONE DEPRESSIO	100753	CLASS II-POSITIVE CELLS		T-LY
HREE DIFFERENT METHODS		GROWTH HORMONE HISTOLOG	94645	NES FROM PATIENTS WITH		VITF
RT OF SEVEN CASES HUMAN		GROWTH HORMONE TRH GONA	97438	MAN RETINAL DETACHMENT		VITF
NIZING HORMONE FSH TSH		GROWTH HORMONE/ THE IMPA	94641	N. VARIETY 85-59 CALLUS	**PROLINE**	ACCUMULAT
MA ANTINEOPLASTIC-DRUG		IMMUNOTHERAPY/ A NEW APP	98106	ENDIPITOUS FINDING OF L		AS A PRECU
UTEINIZING HORMONE AND		IN LACTATING AND NONLACTA	94608	CARBOXYLASE EC 4.1.1.31		ENZYME STRU
NIZING HORMONE FSH AND		IN OVARIECTOMIZED PITUITAR	98442	NEURAMINIC ACID GLYCINE		HISTAMINE/
ECTOMY/ INVOLVEMENT OF		IN THE REGULATION OF PLASM	94668	RO ORGAN CULTURE MODEL		HYDROXYPRO
PIN SECRETION AND BASAL		LEVELS DURING DOPAMINE D-1	99849	OF SALSOLA-SODA ENZYMES		INCOMPATIBI
NEOPLASTIC-DRUG/ SERUM		LEVELS IN PATIENTS TREATED	98029	MA-MANSONI HIGHER FREE		LEVELS IN T
LEX AROMATASE ESTROGEN		LUTEINIZING HORMONE FSH/	101561	ULATION AND EFFECT OF L		ON SOMATIC
SE IN THE INFANTILE RAT		LUTEINIZING HORMONE/ SEXU	94639	TED MICE EGG GRANULOMA		OXIDASE EC
MAN EPILEPSY/ CAPILLARY		MEASUREMENT FOR DIAGNOSI	98225	SEQUENCE DATA/ ROLE OF		RESIDUES IN
EWE RADIOIMMUNOASSAY/		MESSENGER RNA CONCENTRATI	101565	R DEMONSTRATION THAT 9		SUBSTANCE
OPES IN IMMUNOREACTIVE		OF RAT BRAIN MONOCLONAL A	94617	CHROMATOGRAPHY OF THE	**PROLINE-CONTAINI**	
MULATION/ INTRACRANIAL		PERFUSION INDUCES INCUBAT	94630	Y-7-METHYLNONANE INSTAR	**PROLONGATION**	DEAT
UTEINIZING HORMONE FSH		PLASMA CONCENTRATIONS/ OP	91827	DIOGRAPHY/ QT INTERVAL		IN T
TENEDIONE TESTOSTERONE		PLASMA LEVELS FOLLICULAR F	99446	ASSESSEMENT FOR AN ORAL	**PROLONGED-RELEAS**	
YLATE DOPAMINE AGONIST		PROGESTERONE OVULATION/ E	101590	LONING OF PORCINE BRAIN	**PROLYL**	ENDOPEPTIDAS
AND SHORT FORMS OF THE		RECEPTOR ON PROLACTIN-IND	95343	PARENTLY IDENTICAL WITH		ENDOPEPTIDAS
/ DYNAMICS OF PULSATILE		RELEASE DURING THE POSTPAR	101553	F THE OXYTOCIN FRAGMENT	**PROLYL-LEUCYL-GL'**	
EFFECTS OF DOPAMINE ON		RELEASE FROM ECTOPIC AND H	94640	A SYNTHESIS/ FAILURE OF	**PROLYLPEPTIDYL**	ISC
OPIN RELEASING HORMONE		REPRODUCTION/ ENDOCRINE C	91836	ROMASTIGOTE METACYCLIC	**PROMASTIGOTE**	LIPC
ATING SCALE STATISTICS/		RESPONSE TO SUBMAXIMAL ST	100727	IGOTE LOGARITHMIC PHASE		MET
EM/ KALLIDIN STIMULATES		SECRETION FROM INDIVIDUAL	94615	YTE PROLIFERATION ASSAY	**PROMASTIGOTES**	CU
IATED CORTISOL ACTH AND		SECRETION IN HUMANS PHAR	98707	INTRAVENOUS DRUG ABUSE	**PROMISCUOUS**	SEX/
S/ GROWTH HORMONE AND		SECRETION IN HYPOPHYSIAL S	94596	FFERENTIATION OF HUMAN	**PROMONOCYTIC**	U93

Figure 2.10
Detail of a page from the April–May 1991 volume of *Biological Abstracts* listing the key word PROLACTIN and related references. The horizontal arrows target one reference of particular interest.

leads, each of which must be investigated. One entry, indicated in Figure 2.10 by horizontal arrows, does appear somewhat promising:

RECEPTOR ON PROLACTIN-IND . . . AND SHORT FORMS OF THE

Although the paper's precise subject is a bit of a mystery, it does seem to be about prolactin receptors. Ever hopeful, we turn to reference no. 95343 as directed and are rewarded with the listing indicated with an arrow in Figure 2.11. (We could also have arrived at this listing by searching under the key word *milk,* by the way, as indicated by the arrows in Figure 2.12.) Not only does *Biological Abstracts* provide us with the complete citation for the paper by Lesueur *et al.,* but it gives us an informative abstract as well, summarizing the contents of the paper (Figure 2.11). These abstracts are grouped by subject, as indicated in Figure 2.13; we might, therefore, also have discovered the Lesueur *et al.* paper by simply scanning all the abstracts grouped under the heading "Genetics and Cytogenetics."

Note that *Biological Abstracts* can be used to advantage in combination with other indexing services. If you locate an appropriate reference in *Science Citation Index,* for example, you can retrieve an abstract of that paper in *Biological Abstracts.*

Biological Abstracts is currently published in 24 bimonthly issues compiled in two volumes per year; each is about two inches thick—which gives you some idea of the tremendous rate at which research articles are now being published (over 100,000 articles each year!).

Searching Computer Databases

Lastly, I should mention the increasing availability of computerized search services. Through these services, you can enter a series of key words, author names, and subjects, and have the computer search its data banks for relevant papers published within whatever time frame you specify. In a matter of seconds, the computer will examine from 10 to 20 years or so of source material and give you a list of all references relevant to the information you provided. *Zoological Record, Science Citation Index,* and genes to be express[illegible] cells. The enhan[illegible] localized to a 200-[illegible] Rsa I restriction fragments, which contains sequence motifs similar to those found in the other T–cell receptor enhancers but not in the immunoglobulin enhancers.

95342. LETSOU, ANTHEA, SHERRY ALEXANDER, KIM ORTH and STEVEN A. WASSERMAN. (Dep. Biochem., Univ. Tex. Southwestern Med. Cent., Dallas, Tex. 75235.) PROC NATL ACAD SCI U S A 88(3): 810–814. 1991. **Genetic and molecular characterization of tube, a *Drosophila* gene maternally required for embryonic dorsoventral polarity.**—Loss of maternal function of the tube gene disrupts in signaling pathway required for pattern formation in *Drosophila*, causing cells throughout the embryo to adopt the fate normally reserved for those at the dorsal surface. Here we demonstrate that tube mutation also have a zygotic effect on pupal morphology and that this phenotype is shared by mutations in Toll and pelle, two genes with apparent intracellular roles in determining dorsoventral polarity. We [illegible]an describe the isolation of a functionally full–length tube cDNA identified in a phenotypic rescue assay assay. The tube mRNA is expressed maximally early in embryogenesis and again late in larval development, corresponding to required periods of tube activity as defined by distinct maternal and zygotic loss–of–function phenotypes in tube mutations. Sequence analysis of the cDNA indicates that the tube protein contains five copies of an eight–residue motif and shares no significant sequence similarity with known proteins. These results suggest that tube represents a class of protein active in signal transduction at two stages of development.

→**95343.** LESUEUR, LAURENCE*, MARC EDERY*, SUDAH ALI, JACQUELINE PALY*, PAUL A. KELLY and JEAN DJIANE*. (Unite d'Endocrinol. Mol., Inst. Natl. de la Recherche Agronomique, 78352 Jouy–en–Josas Cedex, Fr.) PROC NATL ACAD SCI U S A 88(3): 824–828. 1991. **Comparison of long and short forms of the prolactin receptor on prolactin–induced milk protein gene transcription.**—The biological activities of long and short forms of the prolactin receptor have been compared. These two receptors expressed in mammalian cells were shown to bind prolactin with equal high affinity. The ability of these different forms to transduce the hormonal message was estimated by their capacity to stimulate transcription by using the promoter of a milk protein gene fused to the chloramphenicol acetyltransferase (CAT) coding sequences. Experiments were performed in serum–free conditions to avoid the effect of lactogenic factors present in serum. An ≈ 17–fold induction of CAT activity was obtained in the presence of prolactin when the long form of the prolactin receptor was expressed, whereas no induction was observed when the short form was expressed. The present results clearly establish that only the long form of the prolactin receptor is involved in milk protein gene transcription.

95344. LAU, C. K., M. SUBRAMANIAM, K. RASMUSSEN and T. C. SPELSBERG*. (Dep. Biochem. and Mol. Biol., Div. Endocrinol., Mayo Clin., Rochester, Minn. 55905.) PROC NATL ACAD SCI U S A 88(3): 829–833. 1991. **Rapid induction of the c–jun protooncogene in the avian oviduct by the antiestrogen**

Figure 2.11
Detail of a page from *Biological Abstracts* giving complete citation for a paper by Lesueur *et al.,* along with a detailed summary (abstract) of that paper's contents.

Biological Abstracts are all available in computer-driven versions (called ZOOLOGICAL RECORD ONLINE, SCI SEARCH, and BIOSIS PREVIEWS, respectively). Other useful computerized databases are

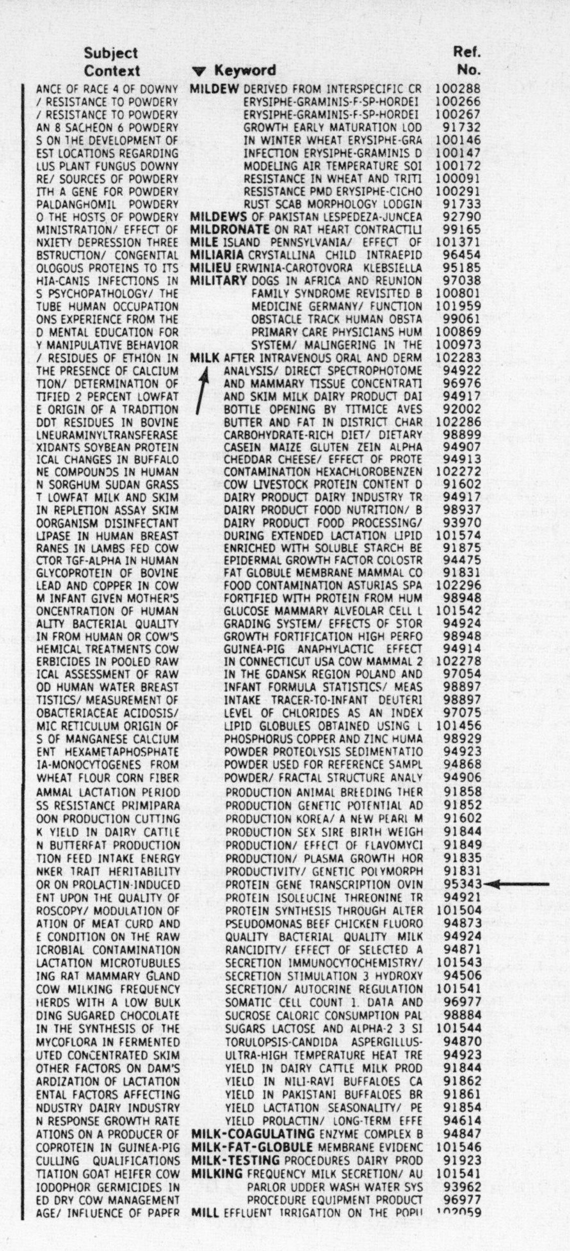

Subject Context	▼ Keyword	Ref. No.
ANCE OF RACE 4 OF DOWNY	**MILDEW** DERIVED FROM INTERSPECIFIC CR	100288
/ RESISTANCE TO POWDERY	ERYSIPHE-GRAMINIS-F-SP-HORDEI	100266
/ RESISTANCE TO POWDERY	ERYSIPHE-GRAMINIS-F-SP-HORDEI	100267
AN 8 SACHEON 6 POWDERY	GROWTH EARLY MATURATION LOD	91732
S ON THE DEVELOPMENT OF	IN WINTER WHEAT ERYSIPHE-GRA	100146
EST LOCATIONS REGARDING	INFECTION ERYSIPHE-GRAMINIS D	100147
LUS PLANT FUNGUS DOWNY	MODELING AIR TEMPERATURE SOI	100172
RE/ SOURCES OF POWDERY	RESISTANCE IN WHEAT AND TRITI	100091
ITH A GENE FOR POWDERY	RESISTANCE PMD ERYSIPHE-CICHO	100291
PALDANGHOMIL POWDERY	RUST SCAB MORPHOLOGY LODGIN	91733
O THE HOSTS OF POWDERY	**MILDEWS** OF PAKISTAN LESPEDEZA-JUNCEA	92790
MINISTRATION/ EFFECT OF	**MILDRONATE** ON RAT HEART CONTRACTILI	99165
NXIETY DEPRESSION THREE	**MILE** ISLAND PENNSYLVANIA/ EFFECT OF	101371
BSTRUCTION/ CONGENITAL	**MILIARIA** CRYSTALLINA CHILD INTRAEPID	96454
OLOGOUS PROTEINS TO ITS	**MILIEU** ERWINIA-CAROTOVORA KLEBSIELLA	95185
HIA-CANIS INFECTIONS IN	**MILITARY** DOGS IN AFRICA AND REUNION	97038
S PSYCHOPATHOLOGY/ THE	FAMILY SYNDROME REVISITED B	100801
TUBE HUMAN OCCUPATION	MEDICINE GERMANY/ FUNCTION	101959
ONS EXPERIENCE FROM THE	OBSTACLE TRACK HUMAN OBSTA	99061
D MENTAL EDUCATION FOR	PRIMARY CARE PHYSICIANS HUM	100869
Y MANIPULATIVE BEHAVIOR	SYSTEM/ MALINGERING IN THE	100973
/ RESIDUES OF ETHION IN	**MILK** AFTER INTRAVENOUS ORAL AND DERM	102283
THE PRESENCE OF CALCIUM	ANALYSIS/ DIRECT SPECTROPHOTOME	94922
TION/ DETERMINATION OF	AND MAMMARY TISSUE CONCENTRATI	96976
TIFIED 2 PERCENT LOWFAT	AND SKIM MILK DAIRY PRODUCT DAI	94917
E ORIGIN OF A TRADITION	BOTTLE OPENING BY TITMICE AVES	92002
DDT RESIDUES IN BOVINE	BUTTER AND FAT IN DISTRICT CHAR	102286
LNEURAMINYLTRANSFERASE	CARBOHYDRATE-RICH DIET/ DIETARY	98899
XIDANTS SOYBEAN PROTEIN	CASEIN MAIZE GLUTEN ZEIN ALPHA	94907
ICAL CHANGES IN BUFFALO	CHEDDAR CHEESE/ EFFECT OF PROTE	94913
NE COMPOUNDS IN HUMAN	CONTAMINATION HEXACHLOROBENZEN	102272
N SORGHUM SUDAN GRASS	COW LIVESTOCK PROTEIN CONTENT D	91602
T LOWFAT MILK AND SKIM	DAIRY PRODUCT DAIRY INDUSTRY TR	94917
IN REPLETION ASSAY SKIM	DAIRY PRODUCT FOOD NUTRITION/ B	98937
OORGANISM DISINFECTANT	DAIRY PRODUCT FOOD PROCESSING/	93970
LIPASE IN HUMAN BREAST	DURING EXTENDED LACTATION LIPID	101574
RANES IN LAMBS FED COW	ENRICHED WITH SOLUBLE STARCH BE	91875
CTOR TGF-ALPHA IN HUMAN	EPIDERMAL GROWTH FACTOR COLOSTR	94475
GLYCOPROTEIN OF BOVINE	FAT GLOBULE MEMBRANE MAMMAL CO	91831
LEAD AND COPPER IN COW	FOOD CONTAMINATION ASTURIAS SPA	102296
M INFANT GIVEN MOTHER'S	FORTIFIED WITH PROTEIN FROM HUM	98948
ONCENTRATION OF HUMAN	GLUCOSE MAMMARY ALVEOLAR CELL L	101542
ALITY BACTERIAL QUALITY	GRADING SYSTEM/ EFFECTS OF STOR	94924
IN FROM HUMAN OR COW'S	GROWTH FORTIFICATION HIGH PERFO	98948
HEMICAL TREATMENTS COW	GUINEA-PIG ANAPHYLACTIC EFFECT	94914
ERBICIDES IN POOLED RAW	IN CONNECTICUT USA COW MAMMAL 2	102278
ICAL ASSESSMENT OF RAW	IN THE GDANSK REGION POLAND AND	97054
OD HUMAN WATER BREAST	INFANT FORMULA STATISTICS/ MEAS	98897
TISTICS/ MEASUREMENT OF	INTAKE TRACER-TO-INFANT DEUTERI	98897
OBACTERIACEAE ACIDOSIS/	LEVEL OF CHLORIDES AS AN INDEX	97075
MIC RETICULUM ORIGIN OF	LIPID GLOBULES OBTAINED USING L	101456
S OF MANGANESE CALCIUM	PHOSPHORUS COPPER AND ZINC HUMA	98929
ENT HEXAMETAPHOSPHATE	POWDER PROTEOLYSIS SEDIMENTATIO	94923
IA-MONOCYTOGENES FROM	POWDER USED FOR REFERENCE SAMPL	94868
WHEAT FLOUR CORN FIBER	POWDER/ FRACTAL STRUCTURE ANALY	94906
AMMAL LACTATION PERIOD	PRODUCTION ANIMAL BREEDING THER	91858
SS RESISTANCE PRIMIPARA	PRODUCTION GENETIC POTENTIAL AD	91852
OON PRODUCTION CUTTING	PRODUCTION KOREA/ A NEW PEARL M	91602
K YIELD IN DAIRY CATTLE	PRODUCTION SEX SIRE BIRTH WEIGH	91844
N BUTTERFAT PRODUCTION	PRODUCTION/ EFFECT OF FLAVOMYCI	91849
TION FEED INTAKE ENERGY	PRODUCTION/ PLASMA GROWTH HOR	91835
NKER TRAIT HERITABILITY	PRODUCTIVITY/ GENETIC POLYMORPH	91831
OR ON PROLACTIN-INDUCED	PROTEIN GENE TRANSCRIPTION OVIN	95343
ENT UPON THE QUALITY OF	PROTEIN ISOLEUCINE THREONINE TR	94921
ROSCOPY/ MODULATION OF	PROTEIN SYNTHESIS THROUGH ALTER	101504
ATION OF MEAT CURD AND	PSEUDOMONAS BEEF CHICKEN FLUORO	94873
E CONDITION ON THE RAW	QUALITY BACTERIAL QUALITY MILK	94924
ICROBIAL CONTAMINATION	RANCIDITY/ EFFECT OF SELECTED A	94871
LACTATION MICROTUBULES	SECRETION IMMUNOCYTOCHEMISTRY/	101543
ING RAT MAMMARY GLAND	SECRETION STIMULATION 3 HYDROXY	94506
COW MILKING FREQUENCY	SECRETION/ AUTOCRINE REGULATION	101541
HERDS WITH A LOW BULK	SOMATIC CELL COUNT 1. DATA AND	96977
DING SUGARED CHOCOLATE	SUCROSE CALORIC CONSUMPTION PAL	98884
IN THE SYNTHESIS OF THE	SUGARS LACTOSE AND ALPHA-2 3 SI	101544
MYCOFLORA IN FERMENTED	TORULOPSIS-CANDIDA ASPERGILLUS-	94870
UTED CONCENTRATED SKIM	ULTRA-HIGH TEMPERATURE HEAT TRE	94923
OTHER FACTORS ON DAM'S	YIELD IN DAIRY CATTLE MILK PROD	91844
ARDIZATION OF LACTATION	YIELD IN NILI-RAVI BUFFALOES CA	91862
ENTAL FACTORS AFFECTING	YIELD IN PAKISTANI BUFFALOES BR	91861
NDUSTRY DAIRY INDUSTRY	YIELD LACTATION SEASONALITY/ PE	91854
N RESPONSE GROWTH RATE	YIELD PROLACTIN/ LONG-TERM EFFE	94614
ATIONS ON A PRODUCER OF	**MILK-COAGULATING** ENZYME COMPLEX B	94847
COPROTEIN IN GUINEA-PIG	**MILK-FAT-GLOBULE** MEMBRANE EVIDENC	101546
CULLING QUALIFICATIONS	**MILK-TESTING** PROCEDURES DAIRY PROD	91923
TIATION GOAT HEIFER COW	**MILKING** FREQUENCY MILK SECRETION/ AU	101541
IODOPHOR GERMICIDES IN	PARLOR UDDER WASH WATER SYS	93962
ED DRY COW MANAGEMENT	PROCEDURE EQUIPMENT PRODUCT	96977
AGE/ INFLUENCE OF PAPER	**MILL** EFFLUENT IRRIGATION ON THE POPU	102059

Figure 2.12
Detail of page from *Biological Abstracts* showing that the Lesueur *et al.* reference could also have been located by using MILK as the key search word (see arrow).

MAJOR CONCEPT HEADINGS FOR ABSTRACTS

Use this list to locate the Abstracts that correspond to your research interest.

	Page
Aerospace and Underwater Biological Effects	AB-1
Agronomy	AB-1
Allergy	AB-28
Anatomy and Histology, General and Comparative	AB-31
Animal Production (includes Fur-Bearing Animals)	AB-32
Bacteriology, General and Systematic	AB-45
Behavioral Biology	AB-46
Biochemistry	AB-54
Biophysics	AB-79
Blood, Blood-Forming Organs and Body Fluids	AB-87
Bones, Joints, Fasciae, Connective and Adipose Tissue	AB-104
Botany, General and Systematic	AB-122
Cardiovascular System	AB-133
Chemotherapy	AB-182
Chordata, General and Systematic Zoology	AB-199
Circadian Rhythm and Other Periodic Cycles	AB-209
Cytology and Cytochemistry	AB-209
Dental and Oral Biology	AB-219
Developmental Biology-Embryology	AB-224
Digestive System	AB-236
Disinfection, Disinfectants and Sterilization	AB-260
Ecology (Environmental Biology)	AB-262
Economic Botany	AB-296
Economic Entomology (includes Chelicerata)	AB-297
Endocrine System	AB-308
Enzymes	AB-336
Evolution	AB-347
Food and Industrial Microbiology	AB-348
Food Technology (non-toxic studies)	AB-359
Forestry and Forest Products	AB-373
General Biology	AB-379
Genetics of Bacteria and Viruses	AB-380
⟶ Genetics and Cytogenetics	AB-400
Gerontology	AB-453
Horticulture	AB-453
Immunology (Immunochemistry)	AB-466
Immunology, Parasitological	AB-521
Integumentary System	AB-526
Invertebrata, Comparative and Experimental Studies	AB-532
Invertebrata, General and Systematic Zoology	AB-554
Laboratory Animals	AB-575
Mathematical Biology and Statistical Methods	AB-575
Medical and Clinical Microbiology (includes Veterinary)	AB-575
Metabolism	AB-604
Methods, Materials and Apparatus, General	AB-612
Microbiological Apparatus, Methods and Media	AB-613
Microorganisms, General (includes Protista)	AB-614*
Morphology and Anatomy of Plants (includes Embryology)	AB-614
Morphology and Cytology of Bacteria	AB-615
Muscle	AB-615
Neoplasms and Neoplastic Agents	AB-625
Nervous System (excludes Sense Organs)	AB-707
Nutrition	AB-791
Paleobiology	AB-801
Paleobotany	AB-802
Paleozoology	AB-803
Palynology	AB-805
Parasitology (includes Ecto- and Endoparasites)	AB-805
Pathology, General and Miscellaneous	AB-811
Pediatrics	AB-812
Pest Control, General (includes Plants and Animals); Pesticides; Herbicides	AB-813
Pharmacognosy and Pharmaceutical Botany	AB-815
Pharmacology	AB-818
Physical Anthropology: Ethnobiology	AB-906
Physiology and Biochemistry of Bacteria	AB-906
Physiology, General and Miscellaneous	AB-919
Phytopathology	AB-922
Plant Physiology, Biochemistry and Biophysics	AB-944
Poultry Production	AB-969
Psychiatry	AB-972
Public Health	AB-995
Radiation Biology	AB-1049
Reproductive System	AB-1057
Respiratory System	AB-1075
Sense Organs, Associated Structures and Functions	AB-1094
Social Biology (includes Human Ecology)	AB-1118
Soil Microbiology	AB-1119
Soil Science	AB-1122
Temperature: Its Measurement, Effects and Regulation	AB-1132
Tissue Culture: Apparatus, Methods and Media	AB-1134
Toxicology	AB-1134
Urinary System and External Secretions	AB-1172
Veterinary Science	AB-1184
Virology, General	AB-1185

*There are no references listed under this heading. See page listed for instructions on locating related topics.

Figure 2.13
Reproduction of a page from the beginning of *Biological Abstracts* showing organization of abstracts. The arrow indicates the topic containing the reference by Lesueur *et al.* (1991).

ENVIROLINE, which covers environmentally related publications

OCEANIC ABSTRACTS, which focuses on marine-related topics, including the effects of pollution

MEDLINE, which covers virtually every aspect of biomedical research

SCISEARCH, which covers every area of pure and applied sciences, including all records published in *Science Citation Index*

TOXLINE, which focuses on the toxicological literature

INDEX TO SCIENTIFIC REVIEWS, which indexes more than 30,000 newly published review articles each year

INDEX TO SCIENTIFIC BOOK CONTENTS, which indexes the individual chapters of multiauthored scientific books

Many libraries subscribe to a service called DIALOG, which covers nearly 400 different computerized databases in a variety of fields. Through DIALOG, the user has access to ZOOLOGICAL RECORD ONLINE, SCI SEARCH, BIOSIS PREVIEWS, MEDLINE, ENVIROLINE, and numerous other databases such as POLLUTION ABSTRACTS and OCEANIC ABSTRACTS. Check with your instructor or reference librarian to find out what relevant databases are available at your college or university.

Computer searches typically cost between about $25 and $45. In addition, you will probably need specialized training before you will be able to search effectively. For these reasons, they are usually more appropriate for graduate students and professionals than for undergraduates. In addition, computer databases rarely include references to papers published before 1970; in most areas of biology, the older literature is a valuable and important resource that should not be overlooked. Thus, your searches should never be limited to computer-based services.

Prowling the Internet

Although there are many ways to cruise the information superhighway, venturing out into the World Wide Web is probably the easiest. There is a phenomenal amount of information out there in the ether, available from something like 200,000 individual Web sites. This abundance creates both opportunities and problems. Prowling the Web can eat up your time the way a vacuum cleaner sucks up dirt. And that is a rather good analogy because much of what is on the Web is junk. Information available in formal scientific journals has gone through a rigorous peer review process; other scientists have evaluated the information that is ultimately presented, and in some cases kept it from being published. Information available through the Internet has typically not been checked by anyone for accuracy. If I wish to announce that snails can fly, no one can stop me. Some Web sites are well worth visiting, but be very careful with the information you obtain. The most reliable information can be found through recognized scientific societies, many of which have created home pages on the Web, and through the growing number of online journals. Papers published in online journals are subjected to the same review procedures that are applied to papers published in hard-copy journals. Internet publishing is a very new venture; we are likely to see more of it in the next few years, particularly once publishers determine how to pay for it.

Using *Current Contents*

One additional search tool of value particularly to postgraduate Biology students is *Current Contents,* published by the Institute for Scientific Information in Philadelphia. *Current Contents* essentially puts the best library in the world at your fingertips. Each weekly issue includes the complete table of contents for scientific journals published a few weeks earlier; more than 950 different journals are covered by the publication, so you are not likely to miss much of the relevant literature in your field, no matter how meager the holdings in your institution's library. If you encounter a paper of particular interest while browsing through the latest issue of *Current Contents,* you can try to find that article in your library or, failing that, you can request a copy of the paper from its author; mailing addresses are given at the back of each issue. Also toward the back of each issue is a subject index, allowing you to locate papers concerning particular topics very quickly. A computerized version of *Current Contents* that includes abstracts of each article referenced is now available.

CLOSING THOUGHTS

Printed indexing services, *Current Contents,* and computerized databases are all good sources of references, but there is a major catch: your library will probably not subscribe to all the journals (or even most of the journals) included in the literature searched by the various services. Although *Current Contents* provides a partial solution to this problem for very recent literature, you must wait from one to several weeks or more before having the actual research paper in hand. You may thus spend considerable time accumulating a long list of intriguing references, creating the comforting illusion that you are getting something done, only to discover that most of the listed journals are not to be found on your campus or in any other nearby library. Consulting recent issues of available, appropriate journals may thus be the most efficient way to search for promising research topics and references.

SUMMARY

1. Become a brain-on reader: work to understand your sources fully, sentence by sentence, figure by figure, and table by table.
2. Try to take notes thoughtfully, in your own words.
3. In your note-taking, be careful to distinguish your words and thoughts from those of the author(s).
4. Be efficient in exploring the primary scientific literature: browse the list of references given in your textbook and in other relevant books, and the papers published in recent issues of relevant scientific journals.
5. Become familiar with the major abstracting and indexing services, including computerized databases, and use these as necessary to complete your literature search. Be cautious about the validity of information acquired on the Internet.

3 *Writing Laboratory Reports*

WHY ARE YOU DOING THIS?

It is no accident that most Biology courses include laboratory components in addition to lecture sessions. Doing Biology involves asking questions, formulating hypotheses, devising experiments to test the hypotheses, presenting data, evaluating data, and interpreting data. Those so-called facts you learn from lectures and textbooks are primarily interpretations of data. By participating in the acquisition and interpretation of data, you glimpse the true nature of the scientific process.

If you are contemplating a career in research, be assured that learning to write effective laboratory reports now is an investment in your future. As a laboratory technician or research assistant, you will often be asked to work up and graph data so that the future path of the research can be decided upon. If you eventually pursue a research M.S. or Ph.D., you will find that a graduate thesis is essentially a large lab report. Writing up your research for publication, as a graduate student or as a researcher with a laboratory of your own, you will quickly find that you are again following the procedures you used in preparing good laboratory reports in college Biology courses; it can't hurt to learn the tools of your trade now.

You can also benefit from writing good laboratory reports even if you do not expect to go on in Biology. Preparing laboratory reports develops the ability to organize ideas logically, think clearly, and express yourself accurately and concisely. It is difficult to imagine a career in which mastery of such skills is not a great asset.

ORGANIZING A LABORATORY OR FIELD NOTEBOOK

The first step in preparing a good laboratory report is to keep a detailed notebook. Laboratory notebooks function to

1. Record the design and goals of your experiments and observations
2. Record and organize your thoughts and questions about the work you are doing or planning to do
3. Record your observations and numerical data
4. Help you organize your activities in the laboratory so that you can work quickly and accurately

Keeping a detailed notebook will make the task of writing your report much easier, and you will end up producing a better report; after all, any product benefits from the use of quality materials. Moreover, the skills you learn in keeping the notebook could really pay off in later life. The Nobel Prize for the isolation of insulin would probably have gone to J. B. Collip instead of Frederick Banting and C. H. Best had Dr. Collip learned to keep a more careful record of his work as a student; Collip was apparently the first to purify the hormone, but his notes were incomplete and he was therefore unable to repeat the procedure successfully in subsequent studies. Don't let something like this happen to you!

In keeping your laboratory notebook, assume that you are doing something worthwhile, that you might well discover something remarkable (it does happen), and that you will suffer complete amnesia while you sleep that night. In other words, take the time to write down—in your own words—everything you are about to do, everything you actually do, and why you do it.

Record your data clearly, with each number identified by the appropriate units. Many of the details that seem too obvious to write down (the name of the species you are working with or the units of measurements, for example) are forgotten surprisingly quickly upon leaving the laboratory; you *can* write down too little, but it's difficult to write down too much.

Write so legibly and clearly that, should you be run over by a truck on your way to class the next day, other students in the course would have no difficulty reconstructing your study and following your results. Similarly, if you use abbreviations in your notebook, be sure to indicate what each stands for. These procedures will greatly facilitate the writing of your laboratory report later, and are in fact crucial in any functioning research laboratory since anyone in the lab must be able to pick up your work or interpret your work where you left off, if you are unable to come in one day or if you leave that laboratory permanently.

Some of this writing can be done before the laboratory session. Whenever possible, read about the day's study ahead of time, being sure (through writing in your notebook) that you understand its goals, and plan exactly how you will record your data. You will get more out of the exercise and will undoubtedly finish your work in the lab sooner if you arrive prepared. It often helps to sketch a simple flowchart of your planned activities ahead of time, as in Figure 3.1.

Your laboratory notebook should contain any thoughts, observations, or questions you have about what you are doing along with the actual protocol and data. A sample page from a model notebook is shown in Figure 3.2. Note that the notebook entry begins by stating the date and a specific goal. With such clear, well-organized, and well thought-out notes, this student is well on his or her way to preparing a fine laboratory report.

A notebook recording field observations (a "field notebook") would look similar to that shown in Figure 3.3, which records part of a study investigating the size distributions of the common marine intertidal snail *Littorina littorea* along a Massachusetts beach.

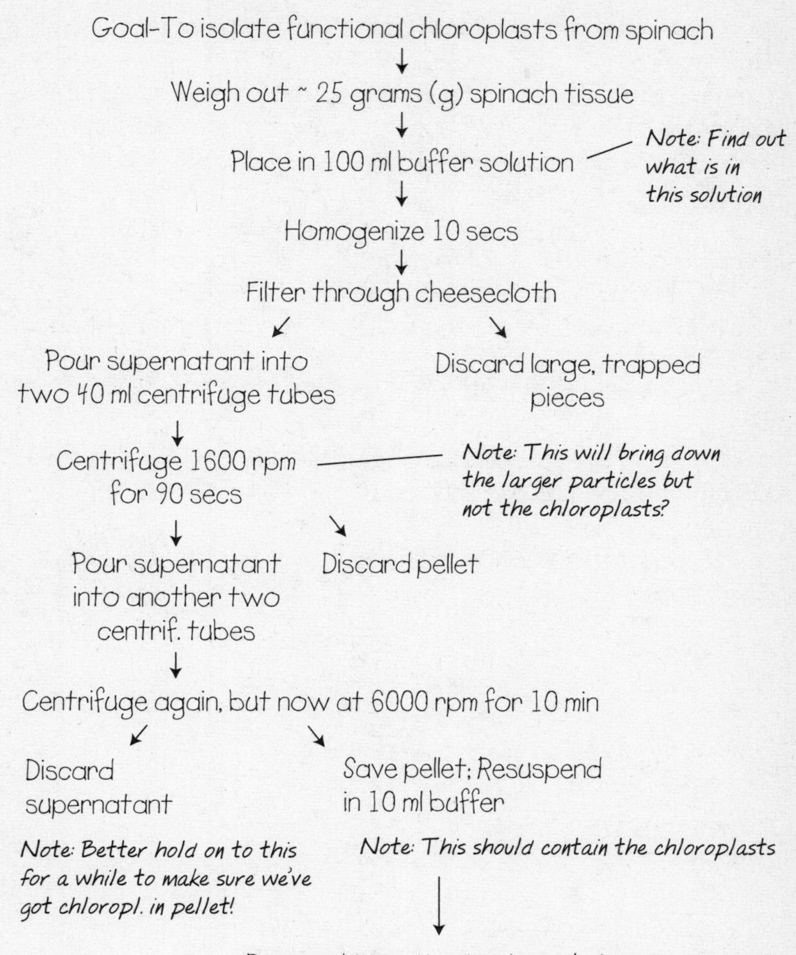

Figure 3.1
A page from a student's laboratory notebook. The flowchart is based on more detailed information provided by the instructor.

October 22, 1996

Goal–to determine rate at which the marine bivalve *Mercenaria mercenaria* (hardshell clam) moves water across its gills.

Approach--Use unicellular alga (phytoplankton) *Dunaliella tertiolecta.*

Determine initial cells $\cdot$ ml $^{-1}$, final cells $\cdot$ ml $^{-1}$. If know elapsed time and volume of seawater in container, can calculate cells eaten per hour per clam. and ml of sea H_2O cleared of cells/h/clam.

Weight of clam (incl. shell): 9.4g [Fisher/Ainsworth balance, Model MX-200]

Find initial concentration of algal cells:

1. 1 ml of culture + drop of Lugol's iodine to kill cells.
2. Load hemacytometer for cell counts--finally got it right on 5th try!

   Helps to tilt pipet at about 45° angle.

(↗ because of dilution w/0.05ml Lugol's sol'n)

Data: (multiply by 1.05 and then by 10^4 -----→ cells ml $^{-1}$)

Note: ask Prof. Scully why 10^4.

counts per section of hemacytom slide

	Sample 1	Sample 2
	22	18
2)	13	21
	18	20
4)	22	20
	26	16
6)	21	12
	20	18
8)	16	22
	22	19
	$\overline{X}$ = 20.0 (x 1.05) ⟶	18.4 (x 1.05) ⟶
	⟶ 21.0 x 10^4 cells ml $^{-1}$	19.4 x 10^4 cells ml $^{-1}$

good agreement!

Put clam in 150ml of this solution at 2:10 PM; H_2O = 21.4°C

Figure 3.2

A sample page from a student's laboratory notebook.

September 17, 1996. 1-3 PM. Sunny, 27° C air temp.; 17° C water.
Goal: investigate size distribution of L. littorea shells at different distances from high tide line.
ME: might expect shells to get larger as get closer to water, since these are marine snails and so should be able to feed more hours per day if spend more time under water. (What do they eat? Can they only eat in water?)

- - - -

Note: Many of the empty shells have large round holes made by a drilling predator--probably Nucella lapillus or Lunatia heros, which seem to be the only carnivorous gastropods living here.
⟶ Should calc. % of dead snails (empty shells) that have drill holes. Drill holes seem to be in exact same location on each shell. How is this possible?

- - - -

Note: Most of the rocks in this area are almost completely covered by very tiny (young) barnacles. Covering = so dense in some places can't even see the rock surface.
Question: Why no big, adult barnacles on these rocks?
Perhaps all die off during the summer for some reason.
There are many large barnacles on rocks much farther down the beach. Same species? If so, why only die here? Would be interesting to come back and monitor survival of young barnacles here and down the beach a few times during the summer and fall--perhaps once a month?
SAMPLING: 0.25m^2 around each transect point, every 2m from high tide (HT) mark.
DATA: No snails found from 0-3 meters (m) below high tide mark.
D = DRILLED

	Shell lengths (cm)	
Distance from high tide (m)	Live snails	Empty shells
4m	1.6, 1.4, 0.5 1.7, 0.9, 2.1	1.4
6m	1.7, 1.9, 1.1 2.0, 1.8, 1.8	1.8D 1.4 1.4

etc.

Figure 3.3
A sample page from a student's field notebook.

Again, the quality of the entries suggests that this student is not simply trying to "get it over with," or to simply write down textbook facts and paraphrases of lecture notes; the student is clearly looking around and thinking—every instructor's dream.

In both laboratory and field work, it is often necessary to draw what you are seeing, particularly when working with living organisms. Perhaps the most important benefit of such drawing is that, done properly, the act of drawing forces you to look more carefully at the object or organism before you. As you draw, pay special attention to the relative sizes, shapes, and textures of different parts. Is part *A* connected to part *B,* or are they only adjacent to each other? Are the widths of parts *A* and *B* similar, or is one wider than the other? How much wider? Is *A* twice as wide as *B,* or three times as wide? The closer you look, the more you will see.

Draw using hard lead pencil if possible, so that you can modify your drawings easily while you work. If you must use pen, be sure the ink does not smear when wet. Try to figure out what things are as you draw, and label them as you go along. Be sure to indicate the approximate size of what you are drawing, along with a scale bar if appropriate. Most important, make your drawings *big* so that they can accommodate plenty of detail. Most beginning students make their drawings much too small; think big. Even if you are looking at something through a microscope, your drawing of that something should fill a 4-by-6-inch space. Remember, the goal is not to become a great artist, but to learn how to observe closely and how to record those observations accurately and in sufficient detail.

COMPONENTS OF THE LABORATORY REPORT

A laboratory report is typically divided into five major sections.

1. *Introduction.* The introductory section, usually only one or two paragraphs long, tells why the study was undertaken; a brief summary of relevant background facts

leads to a statement of the specific problem that is being addressed.

2. *Materials and Methods.* This section is *your* reminder of what you did, and it also serves as a set of instructions for anyone wishing to repeat your study in the future.
3. *Results.* This is the centerpiece of your report. What were the major findings of the study? Present the data or summarize your observations using graphs and tables to reveal any trends you found. Point out these trends to the reader. If you make good use of your tables and graphs, the results can commonly be presented in only one or two paragraphs of text; one picture is worth quite a few words. Avoid interpreting the data in this section.
4. *Discussion.* How do your results relate to the goals of the study, as stated in your introduction, and how do they relate to the results that might have been expected from background information obtained in lectures, textbooks, or outside reading? What new hypotheses might now be formulated, and how might these hypotheses be tested? This section is typically the longest part of the report.
5. *Literature Cited.* This section includes the full citations for any references (including textbooks and laboratory handouts) you have cited in your report. Double-check your sources to be certain they are listed correctly because this list of citations will permit the interested reader to check the accuracy of any factual statements you make, and often, to understand the basis for your interpretations of the data. Cite only material you have actually read.

In some cases, you will be asked to include two additional sections in your reports: the Abstract, in which you summarize the nature of the problem addressed, your approach to the problem, and the major findings and conclusions; and an Acknowledgments section, in which you formally thank people for their contributions to the project.

Before writing your first report, it is helpful to study a few short papers in a major biological journal, such as *Biological Bulletin, Developmental Biology,* or *Ecology.* Reading these journal articles for content is unnecessary; you don't need to understand the topic of a paper to appreciate how the article is crafted. But do pay attention to the way the Introduction is constructed, the amount of detail included in the Materials and Methods section, and the material that is and is not included in the Results section.

In studying an article or two, note that figures and tables are always accompanied by explanatory captions, and that the axes of graphs and the columns and rows of tables are clearly labeled.

WHERE TO START

Strangely enough, the Introduction is not the best place to begin writing your report; it is far easier to write the Introduction toward the end of the job, after you have fully digested what it is that you have done. Start work with either the Materials and Methods section or the Results section. Better still, you may profitably work on the two in tandem; working on the Results section sometimes helps clarify what should be included in the Methods section, and working on the Methods section sometimes clarifies the order in which results should be presented in the Results section.

Because the Materials and Methods section requires the least mental effort, completing it is a good way to overcome inertia. You may not know why you did the experiment or what you found out by doing the experiment, but you can probably reconstruct what you did without much difficulty. Moreover, reminiscing carefully about what you did puts you in the right frame of mind to consider why you did it.

WRITING THE MATERIALS AND METHODS SECTION

Results are meaningful in science only if they can be obtained over and over again, whenever the experiment is repeated. Unfortunately, the results of any study depend to a large extent on the way the study was done. It is therefore essential that you describe your methodology in detail sufficient to permit your experiment to be repeated exactly as originally performed. Perhaps the best reason for writing a detailed Materials and Methods section is that it helps you review what you have done in an organized way and starts you thinking about why you've done it. Developing a good Materials and Methods section puts you in the right frame of mind to do an equally good job on the other sections of the report.

The difficulty in writing this section of a laboratory report (or journal manuscript) is in selecting the right level of detail. Students commonly give too little information; when informed of this defect, they may then give too much information. It's hard to hit it just right, but keeping your audience in mind (yourself and your fellow students) will help.

Determining the Correct Level of Detail

Many students begin with a one-sentence Materials and Methods section: "Methods were as described in the lab manual." Although this sentence meets the criterion of brevity, it is generally unacceptable as a stand-alone Methods section. For one thing, studies are rarely performed exactly as described in a laboratory manual or handout. Your instructions may call for the use of 15 animals, for example, but only 12 animals might be available for use on the day of your experiment. In addition, many details of a study will vary from year to year, week to week, or place to place, and must therefore be omitted from your set of instructions.

But don't get carried away. Consider the following overly detailed description of a study involving the growth of radish seedlings:

> On January 5, I obtained four paper cups, 400 g of potting soil, and 12 radish seeds. I labeled the cups A, B, C, D and planted three seeds per cup, using a plastic spoon to cover each seed with about one-quarter inch of soil.

The author has used the first sentence simply to list the materials; whenever possible, it is far better to mention each new material as you discuss what you did with it. Furthermore, why do we need to know the weight of the soil obtained, or that the cups were labeled A–D rather than 1–4, or that a plastic spoon was used to add soil? Omitting the excess details and starting right in with what was done, we obtain

> On January 5, I planted three radish seeds in each of four individually marked paper cups, covering the seeds with about one-quarter inch of potting soil.

Note that the essential details—individually marked cups, three seeds per cup, one-quarter inch of soil—not only survive in the edited version, but stand out clearly. The trick, then, is to determine which details are essential and which are not.

The best approach to writing the Materials and Methods section is to begin by listing all the factors that might have influenced your results. If, for example, you measured the feeding rates of caterpillars on several different diets, your list might look something like this:

Species of caterpillar used

Diets used

Amount of food provided per caterpillar

Time of year

Time of day

Air temperature in room

Manufacturer and model number of any specialized equipment used (such as balances, centrifuges, or spectrophotometers)

Size and age of caterpillars

Duration of the experiment

Container size

Number of animals per container

Total number of individuals in the study

This list, which you do not turn in with your report, contains the bricks with which you will construct the Materials and Methods section. Each of the listed details must find its way into your report (not necessarily in the order in which you jotted them down) because each gives information essential for later replication of the experiment. Some of this information may also help you explain why your results differed from those of others who have gone before you, a topic that will deserve some attention later, in the Discussion section of your report. Details that do not merit inclusion in the list are superfluous and should not appear in your Materials and Methods section.

In describing the procedures followed, you must say what you did, but you should freely refer to your laboratory manual or handouts in describing how you did it. For example, you might write

```
The three different diets were distributed to the
caterpillars in random fashion, as described in the
laboratory manual (C. Orians, 1996).
```

The important point here is that the diets were distributed at random; the outcome might be quite different if the largest caterpillars were to receive one diet and the smallest caterpillars another. The interested reader, including you, perhaps, at some later date, can refer to the stated source (C. Orians, 1996) for detailed instruction in the method of randomization. You might want to append the relevant portion of your handout or manual at the end of your report as an appendix; this is a fine way to keep everything together for later use.

It is often a good idea to mention, for your own benefit as well as that of your reader, why particular steps were taken. Imagine yourself explaining things to a classmate. We might, for example, profitably rewrite the sentence given in the preceding example to read

> To avoid prejudicing the results by distributing food according to size of caterpillar, the three different diets were distributed to the caterpillars in random fashion as described by Orians (1996).

It is also usually appropriate to include any formulas used in analyzing your data. The following sentences, for example, would belong in a Materials and Methods section:

> The data were analyzed by a series of Chi-Square tests. The rate at which food was eaten was calculated by dividing the weight loss of the food by three hours, according to the following formula: Feeding rate = (Initial food weight − final food weight) ÷ 3 h.

Note that "rates" always have units of "per time." If it doesn't have units of "per time" it is not a rate.

Be sure to note any departures from the given instructions. Suppose you were told to weigh the caterpillars individually but found that your balance was not sensitive enough to record the weight of a single animal. Your laboratory instructor, never at a loss for good ideas, probably suggested that you weigh the individuals in each container as a group. Your report might then include the following information:

> Determining the weight gained by each caterpillar over the three-hour period of the experiment required that both initial and final weights be determined. The caterpillars were too small to be weighed individually. Therefore, similarly sized caterpillars were weighed in groups of three at a time. The average weight of each caterpillar in the group was then calculated.

A Model Materials and Methods Section

The Materials and Methods section of your report should be brief but informative. The following example completely describes an experiment designed to test the influence of decreased salinity on the body weight of a marine worm:

> The polychaete worms used in this study were Nereis virens, freshly collected from Nahant, MA, and ranging in length between 10 and 12 cm. All treatments were performed at room temperature, approximately 21°C, on April 15, 1995. One hundred ml of full-strength seawater was added to each of three 200 ml glass jars; these jars served as controls, to monitor worm weight in the absence of any salinity change. Another three jars were filled with 100 ml of seawater diluted by 50% with distilled water.

Six polychaetes were quickly blotted with paper towels to remove adhering water, and were then weighed to the nearest 0.1 g using a Model MX-200 Fisher/Ainsworth balance. Each worm was then added to one of the jars of seawater. Blotted worm weights were later determined at 30, 60, and 120 minutes after the initial weights were taken.

The initial and final osmotic concentrations of all test solutions were determined using a Model 3W Advanced osmometer, following instructions provided in the handout (G. Lima, 1995).

Note that all essential details have been included: temperature, species used, size of animals used, number of animals used per treatment, number of animals per container, volume of fluid in the containers, type and size of containers, time of year, and equipment used. After reading this Materials and Methods section, you could repeat the study if you wanted to (or had to). Note, too, that the writer has made clear why certain steps were taken; three jars of full-strength seawater served as controls, for example, and worms were blotted dry to remove external water. The fact that worms were blotted dry before they were weighed was mentioned because this is a procedural detail that would obviously influence the results. On the other hand, the author does not describe how the balance was operated, since this technique is standard. The author has written a report that might be useful to someone in the future—and ends up with a top grade.

On to the Results!

WRITING THE RESULTS SECTION

The Results section is the most important part of any research report. Other parts of the report reflect the author's *inter-*

pretation of the data. Interpretations necessarily reflect the author's biases, hopes, and opinions, and are always subject to change, particularly as new information becomes available. In contrast, as long as a study was conducted carefully, and as long as the data were collected carefully and analyzed and presented accurately, the results are valid regardless of how interpretations change over time. Our understanding about how immune systems work has changed remarkably quickly in the past decade, for example, and the current interpretation of older data differs considerably from the original interpretation. But the older data are still valid. The *results* of any study are real; interpretations change. That's why the Results section is indeed the centerpiece of your report.

In this section, you summarize your findings, using tables, graphs, and words. The Results section is

1. Not the place to discuss why the experiment was performed
2. Not the place to discuss how the experiment was performed
3. Not the place to discuss whether the results were expected, unexpected, disappointing, or interesting

Simply present the results, drawing the reader's attention to the major observations and key trends in the data. Don't interpret them here. Most of the work in constructing this section of the report involves data presentation.

Summarizing Data Using Tables and Graphs

Before you even think about doing the writing part of your Results section, you must work with your data. The observations you've made, the data you've collected, most likely contain a story that is crying out for recognition. Contrary to popular opinion, the purpose of tabling and graphing data is not to add bulk to laboratory reports. Rather, you wish to manipulate the data in tables and graphs in order to reveal trends, not only to your instructor but, more importantly, to yourself. The trick now is to organize the data so that (1) the underlying story is revealed and (2) the task of revealing the story to your reader is simplified.

There is no single right way to present summaries of data; use whatever system gives the clearest illustration of trends. You must first decide what relationships might be worth examining, and then experiment with different ways of tabulating and graphing the data to best explore and demonstrate those relationships. Suppose we return to the experiment in which caterpillars fed for three hours on three different diets. We determined both the initial weight of food provided and the weight of food remaining after the three-hour period so that we can calculate the weight of food eaten per caterpillar per hour. In your report, you should provide a sample calculation so that if you make a mistake your instructor can see where you went wrong. We also know the initial and final weights of the caterpillars for each diet, and the initial and final weights of dishes of food in the absence of caterpillars; these control dishes will tell us the amount of water lost by the food because of evaporation.

What relationships in the data might be especially worth examining? The first step in answering this question is to make a list of specific questions that might be worth asking.

1. Do the caterpillars feed at different rates on the different diets? That is, does feeding rate vary with diet?
2. Do larger caterpillars eat faster than smaller caterpillars? That is, does feeding rate vary with size of caterpillar?
3. How is the weight gained by a caterpillar related to the weight lost by the food?
4. Did the weight of the control dishes change, and, if so, by how much?

As in preparing the Materials and Methods section, this list is for your own use and is not to be included in your report. Don't take any shortcuts here. Write these questions in complete sentences. Once you have this list of questions, it is easy to list the relationships that must be examined in your Results section:

1. Feeding rate as a function of diet
2. Feeding rate as a function of caterpillar size
3. Caterpillar weight gain versus food weight loss for each caterpillar
4. Food weight loss in the presence of caterpillars versus food weight loss in controls

Constructing a Summary Table

Now you must organize your data into a table in a way that will let you examine each of these relationships. Consider Table 3.1. This rough draft lists all the data obtained in the experiment; you can work from this table.

For the first relationship in our list (feeding rate as a function of diet), a table will tell the entire story. For your report, you can simply present a summary table, with explanatory caption, as in Table 3.2. Note that one of the caterpillars offered diet A ate no food and lost weight during the experiment. This individual died during the study, and the associated data were therefore omitted from Table 3.2. (The weight loss for this caterpillar probably reflects evaporation of body water.)

To Graph or Not to Graph

Finally, the time has come to reveal more subtle trends that may be lurking in the data. These trends may not be readily apparent from the summary table (Table 3.2); the trends may be made visible, however, to you and your reader, through graphing. A word of caution: do not automatically assume that your data must be graphed. If you can tell your story clearly using only a table, a graph is superfluous. In other cases, you may be able to summarize some aspects of the data without using any graphs or

Table 3.1 Summary of raw data.

Diet	Initial Caterpillar Wt. (g)	Final Caterpillar Wt. (g)	Caterpillar Weight Change (g)	Wt. of Food Lost (g) over 3 h	Feeding Rate (g food lost/h) of Caterpillar
A (Wheatgerm)	8.05	9.55	+1.55	3.65	15.2×10^{-2}
A	4.80	5.80	+1.00	1.74	7.2×10^{-2}
A	5.50	7.00	+1.50	3.33	13.9×10^{-2}
A	5.50	4.70	~~-0.80~~	~~0.00~~	0
A	5.90	6.95	+1.05	1.35	5.6×10^{-2}
Average	5.95	6.80	+1.28	2.52	8.4×10^{-2}
B (Sinigrin	4.40	5.11	+0.71	2.19	9.1×10^{-2}
B 10^{-5}M)	5.20	5.60	+0.40	1.25	5.2×10^{-2}
⋮	⋮	⋮	⋮	⋮	⋮
Control 1	—	—	—	0.22	—
2 (no	—	—	—	0.10	—
3 caterpill.'s)	—	—	—	0.16	—

Table 3.2 The effect of sinigrin (allyl glucosinolate) added to a basic wheatgerm diet, on food consumption of *Manduca sexta* caterpillars over 24 hours.

Diet	No. Caterpillars	(mean g food eaten/ caterpillar/h)
Wheatgerm control	4*	8.4×10^{-2}
Sinigrin (10^{-5} M)	5	7.9×10^{-2}
Sinigrin (10^{-3} M)	5	3.8×10^{-2}

*One individual died during the study without eating any food.

tables. You might write, for example, "No animals ate at temperatures below 15°C," and present data only for animals held at higher temperatures.

Graphs in Biology generally take one of two basic forms: scatter plots (point graphs) or histograms and bar graphs. For the second relationship we wish to examine using the caterpillar data—feeding rate versus caterpillar size—a scatter graph, like Figure 3.4, will be especially appropriate.

In examining Figure 3.4, please note that

1. Each axis of the graph is clearly labeled, including units of measurement.
2. The meaning of each symbol is clearly indicated.
3. A detailed explanatory caption (figure legend) accompanies the figure.

All of your graphs should exhibit these three characteristics. In Figure 3.4, it would be insufficient to simply label the *Y*-axis "Feeding Rate." Feeding rates can be expressed as per minute, per hour, per day, or per year, and can be expressed as per animal, per group of animals, or per gram of body weight. Similarly, it is unacceptable to label the *X*-axis as "Weight," or even as "Caterpillar Weight." Don't make the reader guess what you have done. From the figure caption, the axis labels, and the graph itself, the reader

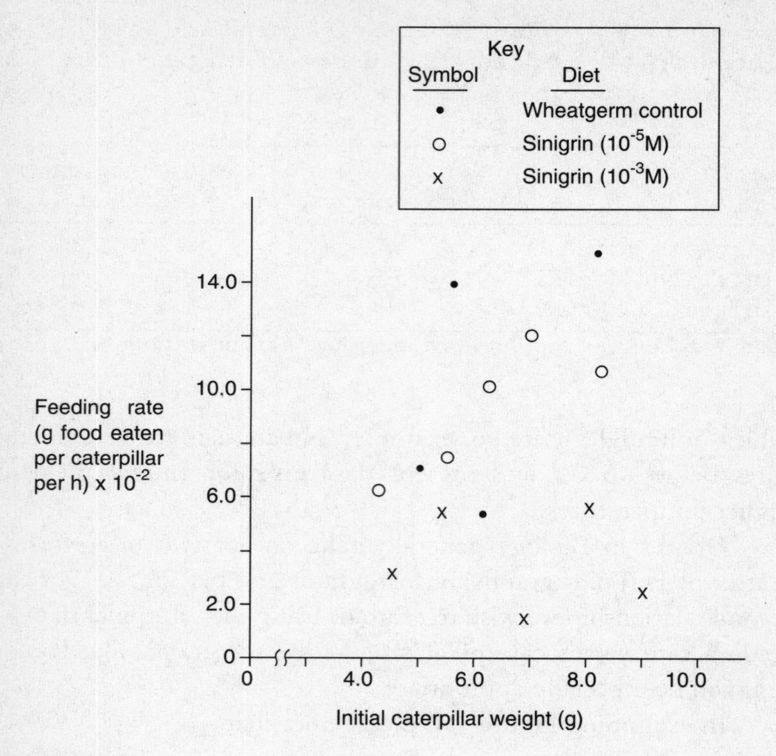

Figure 3.4
The relationship between initial caterpillar weight and rates of food consumption for *Manduca sexta* feeding for 3 hours on one of three different diets at 24°C. Each point represents data from a different individual.

should be able to determine the question being asked, get a good idea of how the study was done, and be able to interpret the figure without reference to the text. Never make the reader back up; a good graph is self-contained.

The third relationship (animal weight gain versus food weight loss) might well be left in table form since in this case the trend is readily discernible; caterpillars always gained less weight than that lost by the food (Table 3.1). The same trend could be revealed more dramatically (or, let us say, more graphically) with a scatter plot, as shown in Figure 3.5, but a graph is not essential here. Again, note the steps taken to avoid ambiguity: the axes are labeled, units of measurement are indicated, symbols are interpreted on the graph, and the figure is accompanied by an informative figure legend. Note also that the symbols used in the student's Figure 3.5 are consistent with their usage in Figure 3.4. Always use the same system of symbols throughout a report so as not to confuse your reader; if filled circles are used to represent data obtained on diet *A* in one graph, filled circles should be used to represent data obtained on diet *A* in all other graphs.

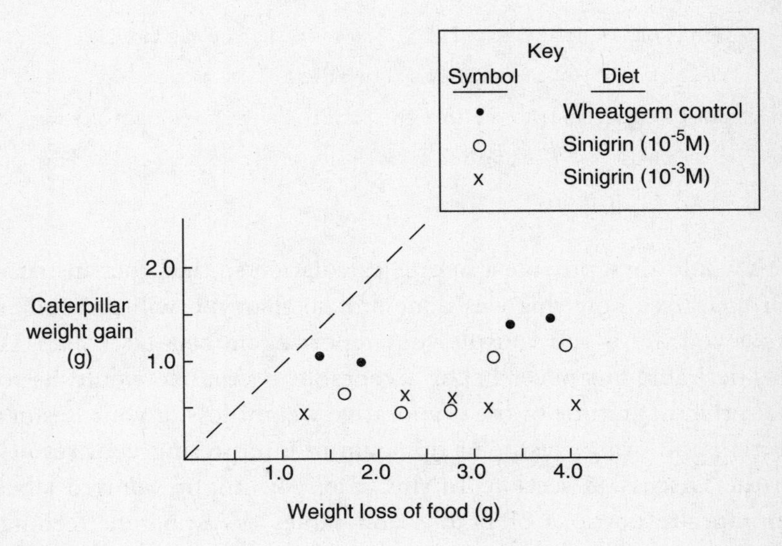

Figure 3.5
Caterpillar weight gain as a function of food consumption for *Manduca sexta* fed for 3 hours on one of three diets at 24°C. Points falling on the dotted line would indicate equality between weight gained and food eaten. Each point represents data from a different individual.

The fourth relationship in our list considers food weight loss in control dishes (no caterpillars). No graphs or tables are needed here; two sentences will do:

```
Control containers exhibited less than a 3% weight
loss (N=3 containers) during the 24 h period. In
contrast, food in containers with caterpillars lost
at least 23% of initial weight.
```

If the weight loss had been substantial, perhaps 5 to 10 percent or more of initial weight, you might wish to adjust all the data in your tables accordingly before making other calculations:

```
Control containers lost 7.6% of their initial
weight (N=3 containers) over the 24 h period. We
therefore adjusted weight loss in other containers
for this 7.6% evaporative loss before calculating
feeding rates.
```

You would then provide a sample calculation so that your instructor could see how this was done and so that you will remember what you did if you consult your report again at a later date. A less desirable but nevertheless acceptable alternative would be to state the magnitude of the evaporative weight loss in your Results section and bring this point up again in interpreting your results in the Discussion section. In this case, you might want to label appropriate portions of graphs and tables as "Apparent feeding rates" rather than "Feeding rates." Again, although there are many wrong ways to present the data, there is no single right way; you must simply be complete, logical, consistent, and clear.

So far, we have looked only at examples of tables and point plots. If you were studying the differences in species composition of insect populations trapped in the light fixtures on four different floors of your Biology building, a bar graph, as in Figure 3.6, might be more suitable. Note again that the axes are clearly labeled, including units of measurement, and that an explanatory legend accompanies the figure. Don't make the reader back up. Note also that the graph tells an interesting story; given that *A* is the fruit fly *Drosophila melanogaster,* it is not difficult to guess where the genetics laboratory is located!

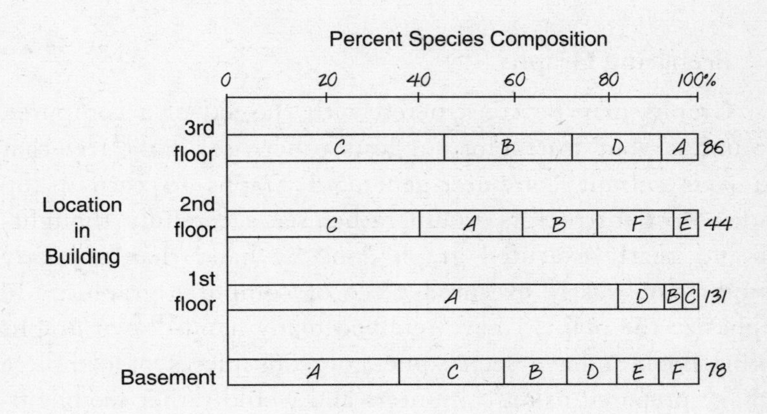

Key

Code	Species
A	*Drosophila melanogaster*
B	*Sarcophaga haemorrhoidalis*
C	*Musca domestica*
D	*Tinea pellionella*
E	unident. sp. A
F	*Malacosoma americanum*

Figure 3.6
The distribution of insect species collected from light fixtures on four floors of the Biology building on May 1, 1996. The number to the right of each bar gives the total number of insects collected on each floor.

Use tables and graphs only if they make your data work for you; if a table or graph fails to help you summarize some trend in your results, it contributes nothing to your report and should be left out. Be selective. Don't include a drawing, graph, or table unless you plan to discuss it, and include only those illustrations that best help you tell your story.

Preparing Graphs

Graphs may be constructed with the aid of a computer, but unless your instructor suggests otherwise, don't feel that you *must* submit computer-generated graphs to earn a top grade. Most instructors would rather see a carefully thought-out and neatly executed graph done by hand than a poorly thought-out, neatly executed piece of computer graphics. To emphasize the point, I have retained many hand-drawn graphs in this book. I have seen some gorgeous pieces of complete garbage prepared using computers and would rather see beginning students spend less time learning to use software and more time thinking about what they present, how they present it, and why they present it.

On the other hand, once you have mastered the key principles of graphing data, learning to use a good software package is certainly worthwhile, particularly since it allows you to quickly examine a variety of relationships in your data and determine which aspects of the data merit graphical presentation. But don't get carried away with all the bells and whistles; once the graphs are plotted, for example, it is sometimes faster to type or hand-print the axis labels or legends than to figure out how to have the computer execute these steps for you.

When preparing graphs by hand, always use graph paper, which can be purchased in your campus bookstore. The most useful sort of graph paper has heavier lines at uniform intervals—at every four-five divisions, for example, as shown in Figures 3.7–3.11. These heavy lines facilitate the plotting of data and reduce eyestrain considerably since every individual line need not be counted in locating data points.

You need not draw directly on the graph paper; it should be perfectly acceptable to place a piece of white tracing paper over a piece of graph paper, trace the *X*- and *Y*-axes, mark off major intervals, and plot points by seeing through the white paper to the underlying grid. In this way, you can use one sheet of graph paper many times. Taping the graph paper to your desk and taping the tracing paper over it will prevent slippage and reduce frustration (and errors).

By convention, the independent variable is plotted on the *X*-axis and the dependent variable is plotted on the *Y*-axis: *Y* is a function of *X*. For example, if you examined feeding rates as a function of temperature (Figure 3.7), you would plot temperature

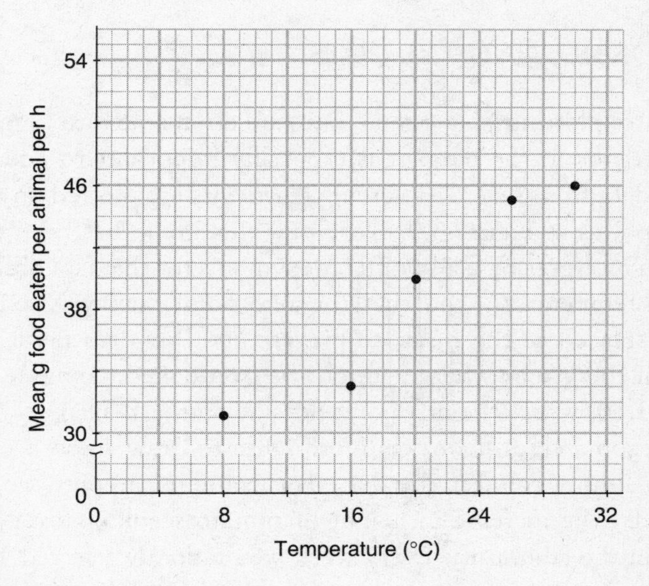

Figure 3.7
Feeding rate of *Manduca sexta* caterpillars on standard wheatgerm diet as a function of environmental temperature. Each point represents the mean feeding rate of 5 individuals measured over 24 hours.

on the *X*-axis and feeding rate on the *Y*-axis; feeding rate *depends* on temperature. On the other hand, temperature is not controlled by feeding rate; that is, temperature varies independently of feeding rate. Temperature is the independent variable and is plotted on the *X*-axis. Note that each point in Figure 3.7 represents data averaged from 5 individuals so that each point is an average, or "mean" value. This is clearly indicated in the *Y*-axis label and in the figure caption.

It is good practice to label the axes of graphs beginning with zero. To avoid generating graphs with lots of empty, wasted space, breaks can be put in along one or both axes, as in Figure 3.7 (see also Figure 3.4 and Figures 3.10–3.12). If a break had not been inserted in the *Y*-axis of Figure 3.7, for example, the graph would have been less compact, as in Figure 3.8.

CONNECTING THE DOTS

After plotting data points, lines are often added to graphs to clarify trends in the data. It is especially important to add such lines if data from several different treatments are plotted on a single graph, as in Figure 3.9. Note that this graph has been made easier to interpret by using different symbols for the data obtained at each temperature, and that the zero point on the *X*-axis has been displaced to the right, to prevent the first data point from lying on the *Y*-axis where it might be overlooked (compare with Figure 3.10, where the first point does lie on the *Y*-axis).

In some cases, it makes more sense to draw smooth curves than to simply connect the dots. For example, suppose we have monitored the increase in height of tomato seedlings over some period in the laboratory. Every week we randomly selected 15-20 seedlings to measure from the laboratory population of several hundred, so that different seedlings were usually measured at each sampling period. After about two months, the data were plotted as in Figure 3.10.

Connecting the dots would not be the most sensible way to reveal trends in the data of Figure 3.10 since we know that the seedlings did not really shrink between days 20 and 28, or between days 36 and 44; simply connecting the points would suggest that shrinkage had occurred. This apparent decline in seedling height reflects the considerable variability in individual growth rates found within the same population, as well as the fact that we did not measure every seedling in the population on every

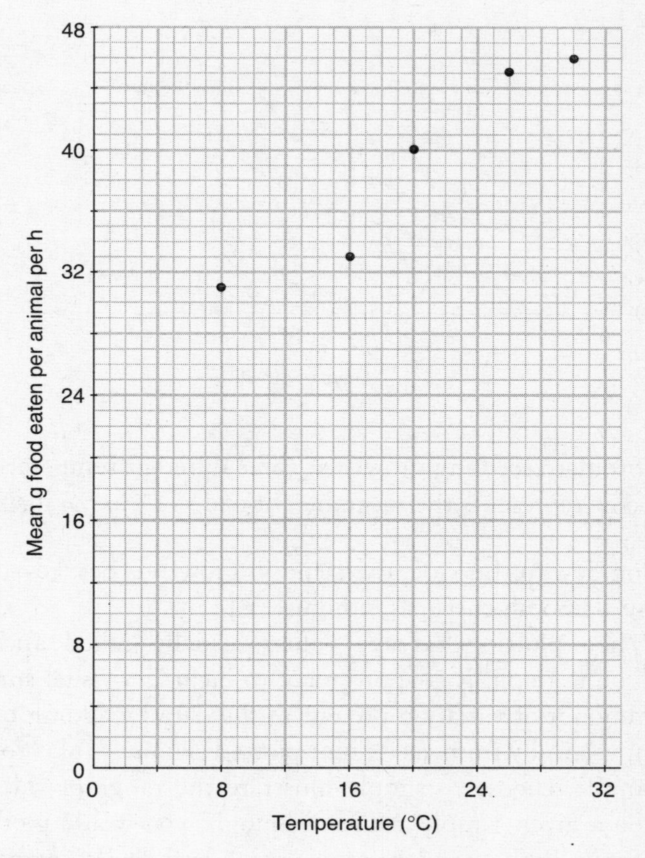

Figure 3.8
Feeding rate of *Manduca sexta* caterpillars on standard wheat-germ diet as a function of environmental temperature. Each point represents the mean feeding rate of 5 individuals measured over 24 hours.

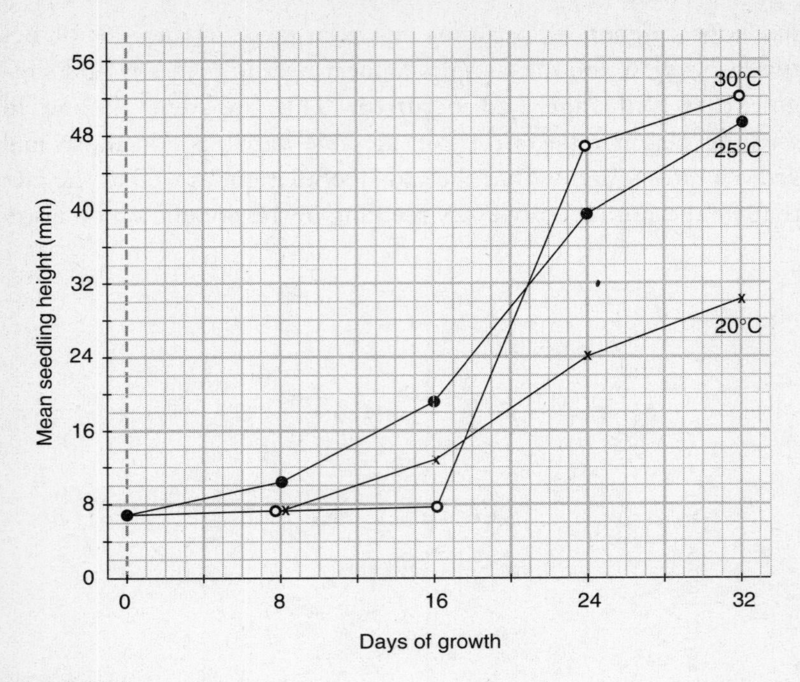

Figure 3.9
Rate of tomato seedling growth at three different temperatures. Each point represents the mean height of 15–17 individuals.

sampling day. In this case, the trend in growth is best revealed by drawing a smooth curve, as in Figure 3.11.

When plotting average values (usually called arithmetic means) on a graph, it is appropriate to include a visual summary of the amount of variation present in the data by adding bars extending vertically from each point plotted (Figure 3.11). You may, for example, choose to simply illustrate the range of values obtained in a given sample. More commonly, you would plot "error bars" (typically the standard error or standard deviation about the mean), giving a visual impression of how much individual data points differed from the calculated mean values, as in Figure 3.11. The less overlap there is between error bars, the more likely it is that differences between mean values are statistically (and biologi-

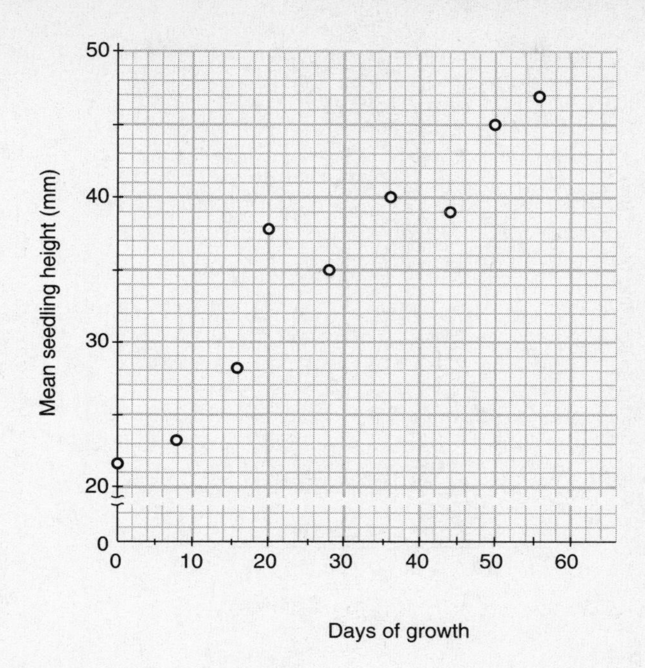

Figure 3.10
Rate of tomato seedling growth at 20°C. Fifteen to 20 seedlings were measured on each day of sampling.

cally) meaningful. Standard deviation and standard error calculations are reviewed in Appendix A.

Plots of standard deviations or standard errors are always symmetrical about the mean value and so convey only partial information about the range of values obtained. If more of your individual values are above the mean than below the mean, the error bars will give a misleading impression about how the data are actually distributed. If your graph is fairly simple, you may be able to achieve the best of both worlds, indicating both the range and standard error (or standard deviation), as in Figure 3.12. In Figure 3.12, the vertical bars extending from the point at day 30 indicate that although the average seedling height was about 37 millimeters (mm) on that day, at least one seedling in the sample was as

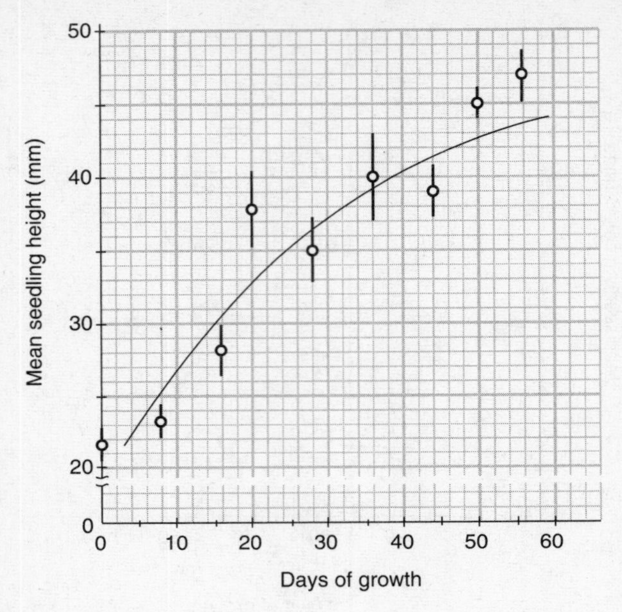

Figure 3.11
Rate of tomato seedling growth at 20°C. On each day of sampling, 15 to 20 seedlings were randomly sampled from a group of 75 seedings and measured. Error bars represent one standard error about the mean.

small as 25 mm and at least one seedling was as large as 48 mm. Seedlings measured on day 50 also differed in height by somewhat more than 20 mm. But we also see that the error bars are much larger for day 30 than for day 50 even though the range of lengths measured was similar for both samples. This tells us that most seedlings were close to the mean length on day 50, about 42 mm long, but that many of the seedlings measured on day 30 were substantially larger or smaller than the mean value measured on that day.

Whether you choose to plot standard deviations, standard errors, ranges, or some other indicator of variation, be sure to indicate in your figure caption what you have plotted, along with the number of points associated with each mean.

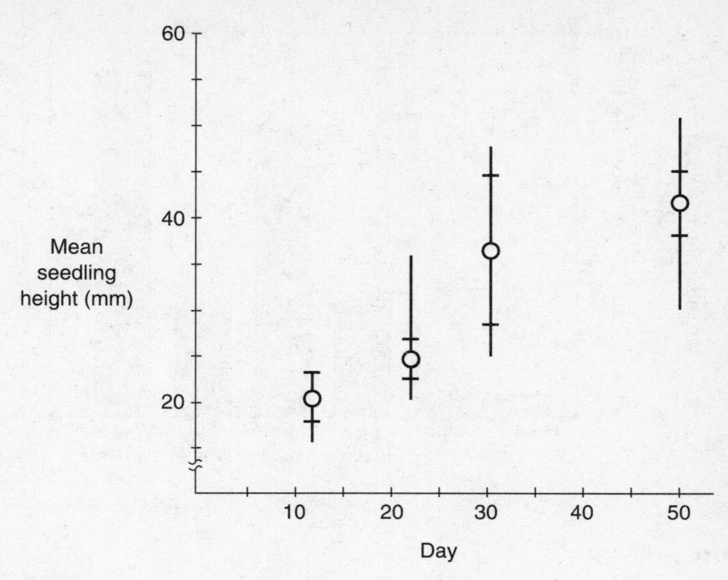

Figure 3.12
Mean height of tomato plant seedlings over a 50-day period. Each point represents the mean height of 18 seedlings raised at 20°C, with a photoperiod of 12L:12D. Vertical bars represent the range of heights; cross-bars represent one standard deviation about the mean.

BAR GRAPHS AND HISTOGRAMS

When the variable along the *X*-axis (the independent variable) is numerical and continuous, points can be plotted and trends can be indicated by lines or curves, as we have seen in Figures 3.4–3.12. In Figure 3.7, for example, the *X*-axis shows temperature rising continuously from 0°C to 32°C, with each centimeter (cm) along the *X*-axis corresponding to a 4°C rise in temperature. Similarly, the *X*-axes of Figures 3.10 and 3.11 reflect the march of time, from 0 to 60 days, with each cm along the *X*-axis reflecting 10 additional days.

When the independent variable is nonnumerical or discontinuous, or represents a range of measurements rather than a single measurement, the data are represented by bars, as shown in

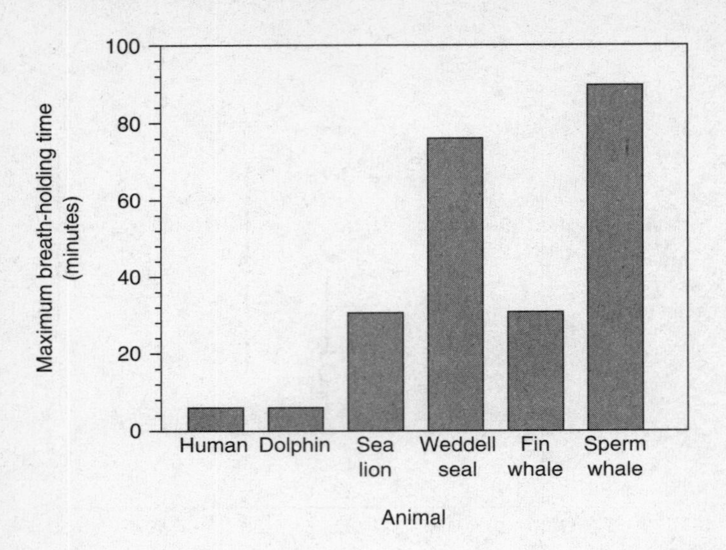

Figure 3.13
Breath-holding abilities of humans and selected marine mammals. Data from Sumich, 1996. *An Introduction to the Biology of Marine Life,* 6th ed. Wm. C. Brown, Publishers.

Figures 3.13 and 3.14. The *X*-axis of Figure 3.13 (a bar graph) is labeled with the names of different mammals. In contrast to the *X*-axes of Figures 3.4–3.12, the *X*-axis of Figure 3.13 does not represent a continuum; no particular quantity continually increases or decreases as one moves along the *X*-axis, and a line connecting the data for sea lion and Weddell seal would be meaningless. In Figure 3.14 (a histogram), the data for shell length are numerical, but are grouped together (for example, all shells 25.0 mm–29.9 mm in length are treated as a single data point). Note also that the magnitude of the size categories represented by the different bars varies; the leftmost bar represents the percentage of shells found within a range of about 0.1 mm to 21 mm in length, whereas each of the next several bars to the right represents the percentage of shells found within a range of only about 5 mm in length. The size range of shells represented by the bar at the ex-

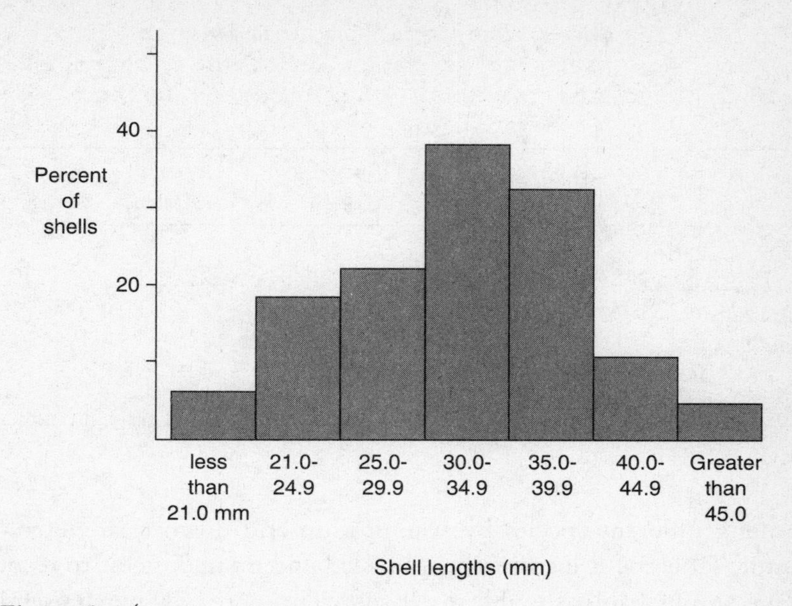

Figure 3.14
Size distribution of snail shells (*Littorina littorea*) collected from the low intertidal zone at Blissful Beach, Massachusetts, on August 15, 1995. Only living animals were measured. A total of 197 snails were included in the survey.

treme right side of the graph is unknown; we know that all shells found in this category exceeded 45 mm in length, but the graph does not indicate the size of the largest shell.

Preparing Tables

Tables should always be organized with data for a given characteristic being presented vertically rather than horizontally. Tables 3.3 and 3.4 present the same information, but in different formats. Table 3.3 correctly places all information about a single species in one row, so that readers can view the information for each species by scanning from left to right, and can compare data

Table 3.3 Characteristics of four snail populations sampled at Nahant, MA, on October 13, 1995. Data are means of 4 replicate samples ± one standard error.

Species	Average Shell Length (cm)	Sample Size	Average No. Animals per m^2
<u>Crepidula fornicata</u>	1.63 ± 0.21	122 indiv.	32.1 ± 4.7
<u>C. plana</u>	1.01 ± 0.34	116	20.8 ± 10.6
<u>Littorina littorea</u>	0.87 ± 0.11	447	113.6 ± 29.1
<u>L. saxatilus</u>	0.40 ± 0.10	60	8.2 ± 5.2

among different species by scanning up and down a single column. Table 3.4 is incorrectly organized and more difficult to read. Like graphs, tables should be self-sufficient; note how much useful information the author has packed into the legend and column headings of Table 3.3.

Table 3.4 Characteristics of four snail populations sampled at Nahant, MA, on October 13, 1995. Data are means of 4 replicate samples ± one standard error.

Species	<u>Crepidula fornicata</u>	<u>C. plana</u>	<u>Littorina littorea</u>	<u>L. saxatilus</u>
Av. shell length (cm)	1.63±0.21	1.01±0.34	0.087±0.11	0.40±0.10
Sample size	122 indiv.	116	447	60
Aver. no. animals per m^2	32.1±4.7	20.8±10.6	113.6±29.1	8.2±5.2

Self-Sufficient Graphs and Tables

A properly executed graph or table is largely self-sufficient: snipped out of your report along with its accompanying legend or caption, it should make perfectly good sense to any Biology major you choose to hand it to. By examining only the axes of a graph and reading the figure caption, for example, the reader should be able to determine the specific question that was asked, how the study was done, and what the main findings are, as discussed in Chapter 2 (pp. 21–25). Consider, for example, Figure 3.11 (p. 82); we can tell quite a lot about how and why this study was done just be carefully studying the figure and its caption. The study was apparently undertaken to determine how fast tomato seedlings grow at one particular temperature, 20°C. We know that the growth of 75 seedlings was followed over nearly 2 months, and that seedlings were measured 9 times over that period, approximately once each week. We also know that not all 75 seedlings were measured every time; instead, the researchers subsampled 15 to 20 each time so that different seedlings were probably measured each time. Finally, we see that each point represents a mean value and that the vertical bars represent one standard error about the mean.

Similarly, from Figure 3.7 (p. 77) we know what species was studied and for how long. We know that caterpillars were maintained at one of five temperatures and we know what those temperatures were; further, we know the number of caterpillars maintained at each temperature.

Try to make your graphs as self-sufficient as these examples are. The easier you make it for readers to understand your data, the more likely it is that your work will be read, and that you will get your point across intact.

Verbalizing Results

One-sentence Results sections are common in student reports: “The results are shown in the following tables and graphs.” However, *common* does not mean *acceptable.* You must use words to draw the reader's attention to the key patterns in your data. But do not simply redraw the graphs in words, as in this description of Figure 3.9.

> At 20°C, the seedlings showed negligible growth for the first 8 days of study. However, between days 8 and 16, the average seedling grew nearly 5 mm, from about 8 mm to about 13 mm. Growth continued over the next 16 days, with the seedlings reaching an average height of 24 mm by day 24, and 30 mm by day 32.

Let the graph do this work for you; your task is to summarize the most important trends displayed by the graph. For example, you might write,

> Temperature had a pronounced effect on seedling growth rates (Figure 3.9). In particular, seedlings at 25°C consistently grew more rapidly than those at 20°C. . . .

Note that the author of this example does not make the reader interpret the data. You must tell the reader exactly what you want him or her to see when looking at your table or graph. Consider the following two examples. The data concern the shell lengths of a particular snail species collected along a rocky coastline, from different regions between the high tide level and the low tide level.

> Although individual specimens of <u>Littorina</u> <u>littorea</u> varied considerably in shell length at each tidal height (Figure xx), there was a significant ($p < 0.05$) distributional effect of shore position on mean size (Table xx).

> Although individual specimens of <u>Littorina</u> <u>littorea</u> varied considerably in shell length at each tidal height (Figure xx), the mean shell length was significantly greater ($p < 0.05$) for snails collected higher up in the intertidal zone (Table xx).

In the first of these examples, the author leaves it to the reader to figure out what specific information is important in the data. Sometimes this is because the author has not taken the time to think about the data carefully enough. The modified version conveys quite a different impression; we know exactly what the author wants us to see, and we can then decide whether we agree with the author's statements or not.

In presenting your own data, first decide exactly what you want your reader to see when looking at each graph or table, and then stick the reader's nose right in it.

Let us apply these principles to the caterpillar study discussed earlier in this chapter. First, is there anything about the general response of the animals worth drawing attention to? You might, for example, be able to write

> All the caterpillars were observed to eat throughout the experiment.

More likely, living things behaving as they do, you will say something like

> One of the animals offered diet <u>A</u> and two of the animals offered diet <u>B</u> were not observed to eat during the three-hour experiment, and the results from these animals were therefore excluded from analysis.

Such a decision to exclude data from further analysis is fine, by the way, as long as you indicate the reason for the decision, and as long as the decision is made objectively; you cannot exclude data simply because it violates a trend that would otherwise be apparent, or because the data contradict a favored hypothesis.

Next, go back to your initial list and reword each question as a statement. For example, the first question posed on p. 68 ("Do the caterpillars feed at different rates on the different diets?"), might be reworded as

```
Caterpillars generally fed at faster rates on the
sinigrin diets than on the wheatgerm controls.
```

In scientific writing, every statement of fact must be backed up with evidence. In this case, you can support the statement with a simple reference to the appropriate figure or table.

```
Caterpillars generally fed at faster rates on the
sinigrin diet than on the wheatgerm controls (Table
3.2).
```

Readers can then look at Table 3.2 (p.71) and decide whether they see the same trend you saw. Sometimes you may want to add a specific example from your data to further support your statement, but that is unnecessary here; one sentence and a table say it all.

If you follow this procedure for each question on your list, your Results section will be complete. The written part will generally be quite short.

Note that the statement about the rate at which caterpillars fed does not mention the term *significant;* it does not say, "Caterpillars fed at significantly faster rates on diet *A* than on diet *B*." Using the term *significant* implies that you have subjected the data to an appropriate statistical test to determine that the differences observed are substantial enough to be convincing. Do not write about significant differences among groups, or the lack of significant differences, unless you have conducted such a test. A brief introduction to the use of statistics in interpreting results begins on

p. 124. This section is worth reading even if you are not required to conduct statistical analyses of your data; as biologists in training, the *how* of statistical analysis is less important to you than the *why.*

Note also the use of the past tense in the statement about caterpillar feeding rates.

```
Caterpillars generally fed at faster rates on diet A.
```

This statement is quite different from the following one, in which the present tense is used:

```
Caterpillars feed at faster rates on diet A.
```

By using the present tense, you would be making a broad generalization extending to all caterpillars, or at least to all caterpillars of the species tested. Before one can make such a broad statement, the experiment must be repeated many, many times, and similar results must be obtained each time; after all, the writer is making a statement about all caterpillars under all conditions. By sticking with the past tense here, you are clearly referring only to the results of your study. Be cautious: always present your results in the past tense.

WRITING ABOUT NUMBERS

According to the Council of Biology Editors, you should use numerals rather than words when writing about counted or measured items, percentages, decimals, magnifications, and abbreviated units of measurement (see pp. 270–271): 6 larvae, 18 seedlings, 25 drops, 25%, 1.5 times greater, 50× magnification, a 3:1 ratio, 0.7 g, 18 ml.

But all rules have exceptions. Use words rather than numerals if beginning a sentence with a number or percentage:

```
Twenty grams of NaCl were added to each of 4 flasks.
```

```
Thirty percent of3 the tadpoles metamorphosed by
the end of the second week.
```

Note the difference in how quantities are presented if we rewrite the first example as, "To each of 4 flasks we added 20 g of NaCl."

When two numbers are written adjacent to each other without being separated by words or a comma, write one of the numbers in words: "The sample was divided into five 25-seedling groups." That sentence could easily be rewritten so that numerals are appropriate for both numbers: "The samples were divided into 5 groups of 25 seedlings each."

When writing about numbers smaller than zero, precede the decimal point with a zero:

```
. . . and we then added 0.25 g NaCl to each flask.
```

When writing about very large or very small numbers, particularly in association with concentrations or rates, use scientific notation. It is preferable, for example, to write about solute concentrations of 5.6×10^{-3} g/ml rather than 0.0056 g/ml, and about cell concentrations being approximately 1.8×10^5 cells/ml rather than about 180,000 cells/ml. Note that I could have avoided both scientific notation and commas in the first example by describing the solute concentration as 5.6 mg/ml. By the way, the word *per,* as in "cells per ml" or "distance per second" may be indicated using either a slash or an exponent: "1.8×10^5 cells/ml" and "1.8×10^5 cells ml^{-1}" are equally acceptable forms of expression.

Finally, the Council of Biology Editors recommends using commas only when numerals contain more than four numbers, as in the following example:

```
Only 1073 of the original 12,450 frog tadpoles died
during the study.
```

Writing About Negative Results

An experiment that was correctly performed always "works." The results may not be what you had expected, but this does not mean that the experiment has been a waste of time. If biologists threw away their data every time something unexpected happened, we would rarely learn anything new. The data you collect are real; it is only the interpretation that is open to question. Therefore always treat your data with respect. The lack of a trend or the presence of a trend contrary to expectation is itself a story worth telling.

IN ANTICIPATION

Much of the work involved in putting together a good laboratory report goes into preparing the Results section. You can save yourself considerable effort and frustration by planning ahead before you enter the laboratory to do the experiment. Be prepared to record your data in a format that will enable you to make your calculations easily. For the caterpillar experiment referred to previously, you would be well ahead by coming to the laboratory with a data sheet set up like the one in Figure 3.15. Using this data sheet, the data are recorded in the *X* areas during the laboratory period; the blank spaces will be filled in later, as you make your calculations. If possible, leave a few blank columns at the right, to accommodate unanticipated needs discovered as you record or work up your data. In introductory laboratory exercises, students are often provided with data sheets already set up in a useful format. It is worth taking a careful look at these data sheets in order to understand how they are organized and why they are as they appear; in more advanced laboratory courses, you will be responsible for organizing your own data sheets. As mentioned earlier, always follow any number you write down with the appropriate units, such as mg (milligrams), cm (centimeters), or mm/min (millimeters per minute, often written as $mm \cdot min^{-1}$). This will avoid potential confusion later.

Date and time started: __________
Date and time ended: __________

Caterpillar No.	Diet	Caterpillar wt. (g)		Weight	Food wt. (g)		Food wt.	Feeding rate
		Initial	Final	Change (g)	Initial	Final	Change (g)	g eaten/caterp./h
X	X	X	X		X	X		

Figure 3.15
Sample format for a laboratory data sheet.

Following the advice of the previous paragraph can save you hours of work later. Even so, it takes time and care to put together an effective Results section. But this section is the heart of your report; craft it properly, and the remainder of the work will be relatively easy.

CITING SOURCES

The next sections to prepare are the Discussion section and the Introduction, in that order. In both sections, you will be making statements of fact that require support, often from written sources. As stated in Chapter 1, every statement of fact or opinion must be supported with a reference to its source. Here are a few general rules to follow when backing up factual statements. These rules apply to all sections of your report.

1. *Don't footnote.* In most papers published in biological journals, references are cited directly in the text, by author and year of publication, as in the following example:

   ```
   A variety of organic molecules are commonly
   used to maintain or adjust the osmotic concen-
   tration of intracellular fluids (Hochachka and
   Somero, 1984; Schmidt-Nielsen, 1990).
   ```

   When more than two authors have collaborated on a single publication, a shortcut is standard practice:

   ```
   A mutation is defined as any change occurring
   in the nitrogenous base sequence of DNA (Tor-
   tora et al., 1982).
   ```

   The *et al.* is an abbreviation for *et alii,* meaning "and others." The words are underlined or italicized, even when abbreviated, because they are in a foreign language, Latin; underlining tells a printer to set the designated words or letters in italics. Note that in each of the examples given, the period follows the closing parenthesis, since the reference, including the publication date, is part of the sentence. Where appropriate, you may incorporate the authors' names directly into a sentence.

```
The various mechanisms known to control gene
activity during embryonic development have been
most recently reviewed by Holliday (1990).
```

or

```
Holliday (1990) has carefully reviewed the lit-
erature concerning control of gene expression
during embryonic development.
```

You can cite your laboratory manual by its author (for example, Professor B. Stewart, 1991) or as follows:

```
Preparation of buffers and other solutions is
described elsewhere (Biology 1 Laboratory Man-
ual, Swarthmore College, 1991).
```

If your information comes from a lecture or from a conversation with a particular individual, support your statement as follows:

```
California gray whales migrate up to 18,000
km yearly (Professor B. Bowen, personal communi-
cation).
```

In some biological journals, the "author–year" format for citing references has been replaced by a more compact "number–sequence" format. In the number–sequence format, each reference cited in the paper is represented by a unique number. The first paper to be cited is assigned the number 1, the second paper to be cited is assigned the number 2, and so forth, as in the following example:

```
A variety of organic molecules are commonly
used to maintain or adjust the osmotic concen-
tration of intracellular fluids (3,5, 8-12).
```

The numerals 3 and 5 represent the 3rd and 5th references to have been cited in the student's paper. Earlier in the paper the author must have cited references 1–7. References 3 and 5, although cited earlier in the paper, are relevant here and so are cited again using the same numbers.

Unless your instructor asks you to adopt this format in your own work, use the author–year format discussed earlier. For one thing, it is a nuisance for the reader to have to keep turning to the back of a paper to see whose work is being cited. Moreover, biologists commonly refer to particular papers by their author(s) and year of publication ("Have you read Wolcott and Wolcott, 1995, yet?"), and using the author–year format of citing references helps you remember who did what. But perhaps most important, research papers are written by real people, and as you become more familiar with the literature in any particular field, you will find yourself coming across many of the same names repeatedly. Using the author–year format is a good way to learn which people are best associated with your field of interest. It is also an excellent way to decide where you might wish to apply for graduate study in the future.

2. *Be concise in citing references.* Avoid writing,

    In his classic work, The Biology of Marine Animals, published in 1967, Colin Nicol reviewed the literature on invertebrate bioluminescence.

    Instead, write,

    The phenomenon of invertebrate bioluminescence has been carefully reviewed by Nicol (1967).

    Again, the period follows the parenthesis.

3. *Cite only those sources you have actually read.* Don't list references simply to add bulk to this section of your report; your instructor is perfectly justified in expecting you to be able to discuss any material you cite. Listing a few references you have thoughtfully incorporated into your paper should do more for your grade than any attempt to create the illusion that you have read everything in the library.

    You may occasionally have to cite a reference that you have not actually read. For example, the results reported by Hendler (1990) may be cited in a book or article written by Dufus (1995), and you have read only the work by Dufus. Your citation should then read, "(Hendler, 1990; as cited by Dufus, 1995)." Let Dufus take the blame if he or she has misinterpreted something.

WRITING THE DISCUSSION SECTION

In this section of the report, you must interpret your results in the context of the specific questions you set out to address in this experiment and in the context of any relevant broader issues that have been raised in lectures, textbook readings, previous coursework, and, possibly, library research. You will consider the following issues:

1. What did you expect to find, and why?
2. How did your results compare with those expected?
3. How might you explain any unexpected results?
4. How might you test these potential explanations?
5. Based on your results, what question or questions might you logically want to ask next?

Clearly, if your results coincide exactly with those expected from prior knowledge, your Discussion section will be rather short. Such a high level of agreement is rarely obtained in three- or four-hour laboratory studies. Indeed, high degrees of variability characterize many aspects of research in Biology, especially at the level of the whole organism. After all, genetically based variation in traits is the raw material of evolution: without such variation, evolution by natural selection would not be possible. Often a study will need to be repeated many times, with very large sample sizes, before clear trends emerge. This point is discussed further in the section on statistical analysis beginning on p. 124. A short paper in a biological journal may well represent years of work by several competent, hardworking individuals. Even the simplest of questions are often not easily answered. Nevertheless, every experiment that was carried out properly tells you *something,* even if that something is not what you specifically intended to find out.

Expectations

State your expectations explicitly, and back your statements up with a reference. Scientific hypotheses are not simply random guesses. Your expectations must be based on facts, not opinions; these facts could come from lectures, laboratory manuals or handouts, textbooks, or any other traceable source. In discussing a study on the effectiveness of different wavelengths of light in promoting photosynthesis, for example, you might write something like

> All wavelengths of light are not equally effective in promoting photosynthesis: green light is said to be especially ineffective (Ellmore and Jones, 1991). This is because green light tends to be reflected rather than absorbed by plant pigments, which is why most plants look green (Ellmore and Jones, 1991). Our results supported this expectation. In particular . . .

Alternatively, a discussion section might profitably begin

> The results of our experiment failed to support the hypothesis (Milburn and Jones, 1981) that caterpillars of <u>Manduca sexta</u> reared on one uniquely flavored diet will prefer that diet when subsequently given a choice of foods.

Here we have managed to state our expectations and compare them with our results in a single sentence. In both cases, we have begun our discussion on firm ground—with facts rather than unsupported opinions.

Explaining Unexpected Results

When results refuse to meet expectations, students commonly blame the equipment, the laboratory instructor, the laboratory partners, or themselves. Generally, more scientifically interesting possibilities than experimenter incompetence are the culprits. Don't be too hard on yourself. Take another look at the list of factors you wrote down when beginning to work on your Materials and Methods section. Could any of these factors be sufficiently different from the normal or standard conditions under which the experiment is performed to account for the difference in results? Look again at your laboratory manual or handout. Are any of the conditions under which your experiment was performed substantially different from those assumed in the instruction manual? If you discover no obvious differences in the experimental conditions, or if the differences cannot account for your results, include this point in your report as in this example:

> The discrepancy in results cannot be explained by the unusually low temperature in the laboratory on the day of the experiment, since the control animals were subjected to the same conditions and yet behaved as expected.

If potentially important differences are noted, put this ammunition to good use.

> In prior years, these experiments have been performed using species X. It is possible that species Y simply behaves differently under the same experimental conditions.

Note that the writer does not *conclude* that species *X* and species *Y* behave differently; the writer merely *suggests* this explanation as a possibility. Always be careful to distinguish possibility from fact. Suggesting a logical possibility won't get you into any trouble. Stating your idea as though it were an accepted fact, on the other hand, is sticking your neck out far enough to get your head chopped off.

Continue your discussion by indicating possible ways that the differences in behavioral responses might be tested. For example:

> This possibility can be examined by simultaneously exposing individuals of both species to the same experimental conditions. If species X behaves as expected, and species Y behaves as it did in the

```
present experiment, then the hypothesis of species-
specific behavioral differences will be supported.
If species X and species Y both respond as species
Y did in the present study, then some other expla-
nation will be called for.
```

Continue in this vein, evaluating all the reasonable, testable possibilities you can think of. An instructor enjoys reading these sorts of analyses because they indicate that students are thinking about what they've done. Go ahead; make an instructor happy.

Notice that in the preceding example the writer did not say, "If species *X* behaves as expected, and species *Y* behaves as it did in the present experiment, then the hypothesis will be supported." This writer remembers rule number 8 (p. 8): Never make the reader back up. Notice, too, that the writer does not write, ". . . then the hypothesis will be true" or ". . . then the hypothesis will be proven." Experiments cannot *prove* anything; they can only support or disagree with hypotheses. As scientists, our interpretations of phenomena may make excellent sense based upon what we know at the moment, but those interpretations are not necessarily correct. New information often changes our interpretations of previously acquired data.

Analysis of Specific Examples

EXAMPLE 1

In this study, tobacco hornworm caterpillars were raised for four days on one diet, and then tested over a three-hour period to see if they preferred that food when given a choice of diets.

STUDENT PRESENTATION

```
    The data indicate that the choice of food was not
related to the food upon which the caterpillars had
```

been reared. These data run counter to the hypothesis (Back and Reese, 1976) that hornworms are conditioned to respond to certain specific foods. Only one set of data out of the four gave any indication of a preference for the original diet, and that indication was rather weak.

There are many possible explanations for data that are so contrary to previous experimental results. Inexperience of the experimenters, combined with the fact that three different people were recording data about the worms, may account for part of the error. Keeping track of many worms and attempting to interpret their actions as having chosen a food or having merely been passing by may have proven to be too much for first-time worm watchers. The mere fact that each of three people will interpret actions differently and will have somewhat different methods of recording information introduces bias into the data.

ANALYSIS

This Discussion section starts out well, with a comparison between results expected and results obtained. The hypothesis being discussed is clearly stated, and a supporting reference is given. The student even recognizes that "data are" rather than "data is." (On the other hand, the student persists in calling the animals worms rather than caterpillars; the term *worms* usually refers to annelids, whereas these animals are arthropods). The next paragraph, however, betrays a total lack of confidence in the data obtained; the results could not possibly have turned out this way unless the researchers were incompetent, writes the student.

Although inexperience can certainly contribute to suspicious results, are there no other possible explanations? Does it really take years of training to determine whether a caterpillar is eating food *A* or food *B?*

Compare this report with the one in the next example. This Discussion section deals with the same experiment. In fact, the two students were laboratory partners.

EXAMPLE 2

Contrary to expectation, the results suggest that caterpillars show no preference in the diet they touched first and the diet they spent the most time feeding on. This unexpected finding may be due to the fact that the caterpillars were not reared on the original diets for a long enough period of time to acquire a lasting preference. They ate only the diets they were reared on for four days, whereas the laboratory handout had suggested a pre-feeding period of 5-10 days (Chew and Lewis, 1994). This possibility may be tested by performing the same experiment but varying the amount of time that the caterpillars are reared on the original diets. Such an experiment would determine whether there is a critical time that caterpillars should be reared on a particular diet before they will show a preference for that diet. Another possible explanation for the results obtained may be that the caterpillars used were very young, weighing only 3-6 mg. Finally, this experiment lasted only 3 hours. Perhaps different results would have

```
been obtained had the organisms been given more
time to adjust to the test conditions. It would be
interesting to run the identical experiment for a
longer period of time, such as 10-12 hours.
```

The author of this report produced a paper that clearly indicates thought. Which report do you suppose received the higher grade?

EXAMPLE 3

In this experiment, several hundred milliliters (ml) of filtered pond water were inoculated with a small population of the ciliated protozoan *Paramecium multimicronucleatum* and then distributed among three small flasks. Over the next five days, changes in the numbers of individuals per ml of water in each flask were monitored.

STUDENT PRESENTATION

```
   The large variation observed between the groups
of three replicate populations suggests that the ex-
perimental technique was imperfect. The sampling er-
ror was high because it was difficult to be precise
in counting the numbers of individuals. Some animals
may have been missed while others were counted re-
peatedly. More accurate data may be obtained if the
number of samples taken is increased, especially at
the higher population densities. In addition, more
than three replicate populations of each treatment
could be established. Finally, extremely precise mi-
croscopes and pipets could be used by experienced
operators in order to reduce sampling error.
```

ANALYSIS

This writer, like the writer of Example 1, starts out by assuming that the experiment was a failure, and spends the rest of the report making excuses for this failure. The quality of the microscopes was certainly adequate to recognize moving objects, and *P. multimicronucleatum* was the only moving organism in the water; the author is grasping at straws. If the author had more confidence in his or her abilities, the paper might have been far different. Isn't there some chance that the experiment was performed correctly? Lacking confidence in the data, the student looked no further even though he or she actually had access to data that would have allowed several of the stated hypotheses to be assessed. On each day, for example, several sets of samples were taken from each flask, and each set gave similar estimates for numbers of organisms per ml; this consistency of results suggests that the variation in population density from flask to flask was not due to experimenter incompetence. In addition, although the student stated correctly that larger sample sizes would have been helpful, he or she should have supported that statement with additional analysis of the data. Fifteen drops were sampled from each flask for each set of samples. The student could have calculated the mean number of individuals in the first 3 drops, the first 6 drops, the first 9 drops, the first 12 drops, and then the full 15 drops, to see how estimates of population size changed as the sample size was increased. If the student had done this, he or she would probably have found that larger sample sizes are especially important when population density is low. (Why might this be so?)

EXAMPLE 4

In this last study, a group of students went seining for fish in a local pond. Every fish was then identified to species so that the number of fish of each species could be determined. It turned out that 91 percent of the fish in the sample of 73 individuals belonged to a single species. The remaining fish were distributed among only two additional species.

STUDENT PRESENTATION

I find the small number of species represented in our sample surprising, since the pond is fed by several streams that might be expected to introduce a variety of different species into it, assuming that the streams are not polluted. The lab manual states that 12 fish species have been found in the adjacent streams. It appears that the conditions in the pond at the time of our sampling were especially suitable for one species in particular of all those that most likely have access to it. Perhaps the physical nature of the pond is such that the number of niches is small, in which case competition would become very keen; only one species can occupy a given niche at any one time (Smith, 1958). The reproductive pattern of the fishes might also contribute to the observed results. Possibly Lepomis macrochirus, the dominant species, lays more eggs than the others, or perhaps the juveniles survive better.

Another possible explanation for our findings might lie in the fact that we sampled only the perimeter of the pond, since our seining was limited to a depth of water not exceeding the heights of the seiners. The species distribution could be very different in the middle of the pond at a greater depth.

ANALYSIS

I have not reproduced the entire Discussion section of the student's paper, but even this excerpt demonstrates that a little thinking goes a long way. Note that the student did not require much specialized knowledge to write this Discussion section, only a bit of confidence in the data. Another student might well have written

```
Most likely, the fish were incorrectly identi-
fied; more species were probably present than
could be recognized by our inexperienced team. It
is also possible that the net had a large tear,
which let most of the species escape. I didn't
notice this rip in the fabric, but my glasses
were probably dirty, and then again, I'm not very
observant.
```

WRITING THE INTRODUCTION SECTION

The Introduction section establishes the framework for the entire report. In this section, you briefly present background information that leads to a clear statement of the specific issue or issues that will be addressed in the remainder of the report; by the time you have completed writing the Materials and Methods, Results, and Discussion sections of your laboratory report, you should be in a good position to know what these issues are. In one or two paragraphs, then, you must present an argument explaining why the study was undertaken. More to the point, perhaps, the Introduction provides you with your first opportunity to convince your instructor that you understand why you have been asked to do the exercise.

Every topic that follows this section should be anticipated clearly in the Introduction. Conversely, the Introduction should contain only information that is directly relevant to the rest of the report.

Stating the Question

Even though the statement of questions posed, or issues addressed, generally concludes the Introduction section of a report, it is helpful in writing this section of the paper to deal with this issue first. What *was* the point of this study?

Write the following words: "In this study" or "In this experiment." Then complete the sentence as specifically as possible. Three examples follow.

> In this study, the oxygen consumption of mice and rats was measured in order to investigate the relationships between metabolic rate, body weight, and body surface area.

> In this study, we collected fish from two local ponds and classified each fish into its proper taxonomic category.

> In this experiment, we asked the following question: Do the larvae of Manduca sexta prefer the diet upon which they have been reared when offered a choice of diets?

Note that each statement is phrased in the past tense since the students are describing studies that have now been completed.

The strong points of these statements are best revealed by examining a few unsatisfactory ways to complete sentences dealing with the same material.

> In this study, we measured the metabolic rate of rats and mice.

```
In this study, we made a variety of measurements
on fish.

In this experiment, the feeding habits of Manduca
sexta larvae were studied.
```

Each of these unsatisfactory statements is vague; the reader will assume, perhaps correctly, that you are as much in the dark about what you've done as your writing implies. Be specific. Here, in one sentence, you must come fully to grips with your experiment or study. There *was* some point to the time that you were asked to spend in the laboratory; find it.

AN ASIDE: STUDIES VERSUS EXPERIMENTS

An experiment always involves manipulating something, such as an organism, an enzyme, or the environment, in a way that will permit specific hypotheses to be tested. Containers of protozoans in pond water could be distributed among three different temperatures, for example, to test the influence of temperature on the reproductive rate of the particular species under study. As another example of an experiment, the ability of salivary amylase to function over a range of pHs might be examined to test the hypothesis that the activity of this enzyme is pH sensitive. In the field, a population of marine snails from one location might be transplanted to another location and the subsequent survival and growth of the transplanted population studied so as to test the hypothesis that conditions in the new location are less hospitable for that species than in the location from which the original population was obtained. As a control, of course, the survival and growth of animals not transplanted would also have to be monitored over the same period. Note that an experiment may be conducted in the laboratory or in the field.

It is permissible to refer to experiments as "studies," but not all studies are "experiments." In contrast to the preceding experiments, some exercises require you to collect, observe, enumerate, or describe. You should avoid referring to such studies as experiments; where there are no manipulations, there are no experiments. You might, for example, collect insects from light fixtures located at several different locations within the Biology building and identify these insects to species, enabling you to examine the distribution of insects within the building. Or you might be asked to provide a detailed description of the feeding activities of an insect. Or you might spend an afternoon documenting the depth to which light penetrates into various areas of a lake, and then correlate that information with data on the distribution of aquatic plants in the different areas. In each case, you should refer to your work as a study, not as an experiment. For example:

> In this study, insects were collected from all light fixtures on floors 1, 3, and 5 of the Dana building, and the distribution of species among the different locations was determined.

Providing the Background

Having posed, in a single sentence, the question or issue that was addressed, it is relatively easy to fill in the background needed to understand why the question was asked. A few general rules should be kept in mind.

1. *Support all statements of fact with a reference to your textbook, laboratory manual, outside reading, or lecture notes.* Unless you are told otherwise by your instructor, do not use footnotes. Rather, refer to your reference within the text, giving the author of the source and the year of publication, as in the following example:

   > Many marine gastropods enclose their fertilized eggs within structurally complex encapsu-

```
lating structures (Hunt, 1966; Tamarin and
Carriker, 1968).
```

Note that the period concluding the sentence comes after the closing parenthesis.

2. *Define specialized terminology.* Most likely, your instructor already knows the meaning of the terms you will use, but by defining them in your own words in your report you can convince the instructor that you, too, know what these words mean. Write to illuminate, not to impress. As always, if you write with your future self in mind as the audience, you will usually come out on top; write an Introduction you will be able to understand five years from now. The following two examples obey this and the preceding rule:

```
A number of caterpillar species are known to
exhibit induction of preference, a phenomenon
in which an organism develops a preference for
the particular flavor on which it has been
reared (Jones and Smith, 1983).

Molluscs are common inhabitants of the inter-
tidal zone, that region of the ocean lying be-
tween the high and low tide marks (Jones and
Smith, 1983). The development of mature female
gametes, a process termed oogenesis, is regu-
lated by changing hormonal levels in the blood
(Gilbert, 1991; Gaudette, 1993).
```

3. *Never set out to prove, verify, or demonstrate the truth of something.* Rather, set out to test, document, or describe. In Biology (and science in general), truth is elusive; it is important to keep an open mind when you begin a study and when you write up the results of that study. It is not

uncommon to repeat someone else's experiment or observations and obtain a different result or description. Responses will differ with species, time of year, and other, often subtle, changes in the conditions under which the study is conducted. To show that you had an open mind when you undertook your study, you would want to revise the following sentences before submitting them to your instructor:

> In this experiment, we attempted to demonstrate induction of preference in larvae of Manduca sexta.

> This study was undertaken to verify the description of feeding behavior given for Manduca sexta by Jones (1903).

> This experiment was designed to show that pepsin, an enzyme promoting protein degradation in the vertebrate stomach, functions best at a pH of 2, as commonly reported (Jones and Smith, 1983).

The first example might be modified to read,

> In this experiment, we tested the hypothesis that young caterpillars of Manduca sexta demonstrate the phenomenon of induction of preference.

How would you modify the other two examples to show that you approached the studies without prejudice?

4. *Be brief.* Include only the information that directly prepares the reader for the statement of intent, which will

appear at the end of the Introduction section as already discussed. If, for example, your study was undertaken to determine which wavelengths of light are most effective in promoting photosynthesis, there is no need to describe the detailed biochemical reactions that characterize photosynthesis. As another example, consider these few sentences taken from a report describing an induction of preference study. Caterpillars were reared on one diet for five days and tested later to see if they chose that food over foods that the caterpillars had never before experienced.

In this experiment, we explored the possibility that larvae of Manduca sexta could be induced to prefer a particular diet when later offered a choice of diets. The results of this experiment are significant because induction of preference is apparently linked to (1) the release of electrophysiological signals by sensory cells in the animal's mouth and (2) the release of particular enzymes, produced during the period of induction, that facilitate the digestion and metabolism of secondary plant compounds (laboratory handout, 1994).

The entire last sentence does not belong in the Introduction. The work referred to in this example was a simple behavioral study; students did not make electrophysiological recordings, nor did they isolate and characterize any enzymes. Although a consideration of these two topics might profitably be incorporated into a discussion of the results obtained, these issues should be excluded from the Introduction because they provide no rationale for this particular study; they do not explain why the study was undertaken. Include in your Introduction section only information that directly prepares the reader for the final statement of intent. You might, on a separate piece of paper, jot down other ideas that occur to you for possible use in revising your Discussion section, but if they don't make a contribution here, don't let them intrude on your Introduction. Be firm. Stay focused.

5. *Write an Introduction for the study that you ended up doing.* Sometimes it is necessary to modify a study for a particular set of conditions so that the observations actually made no longer relate to the questions originally posed in your laboratory handout or laboratory manual. For example, the pH meter might not have been working on the day of your laboratory experience, and your instructor modified the experiment accordingly; perhaps the experiment you actually performed dealt with the influence of temperature, rather than pH, on enzymatic reaction rates. In such an instance, you would make no mention of pH in your Introduction section since the work you ended up doing dealt only with the effects of temperature.

The following paragraphs satisfy all the requirements of a valid introduction. This introduction is brief but complete—and effective:

> It is well known that plants are capable of using sunlight as an energy source for carbon fixation (Ellmore and Smith, 1983). However, all wavelengths of light need not be equally effective in promoting such photosynthesis. Indeed, the green coloration of most leaves suggests that wavelengths of approximately 550 nm are reflected rather than absorbed

so that this wavelength would not be expected to produce much carbon fixation by green plants.

During photosynthesis, oxygen is liberated in proportion to the rate at which carbon dioxide is fixed (Ellmore and Smith, 1983). Thus relative rates of photosynthesis can be determined either by monitoring rates of oxygen production or by monitoring rates of carbon dioxide uptake. In this experiment, we monitored rates of oxygen production under different light conditions to test the hypothesis that wavelengths differ in their ability to promote carbon fixation by Elodea canadensis.

DECIDING ON A TITLE

A good title summarizes, as specifically as possible, what lies within the Introduction and Results sections of the report. Your instructor is a captive audience. In the real world of publications, however, your article will vie for attention with articles written by many other people; the busy potential reader of your paper will often glance at the title of your report and promptly decide whether to stay or move on. The more revealing your title is, the more easily your potential audience can assess the relevance of your paper to their interests. A paper that delivers something other than what is promised by the title can lose you considerable good will when read by the wrong audience and may be overlooked by the audience for which the paper was intended. Indeed, many potential readers will miss your paper entirely since indexing services such as *Biological Abstracts* and *Current Contents* use key words from a paper's title in preparing their subject indexes.

Here is a list of mediocre titles, each followed by one or two more revealing counterparts.

No: Metabolic rate determinations
Yes: Exploring the relationship between body size and oxygen consumption in mice

No: Plankton sampling in Small Pond
Yes: Species composition of the spring zooplankton of Small Pond, MA

No:

1. Measuring the feeding behavior of caterpillars
2. Eating habits of *Manduca sexta*
3. Food preferences of *Manduca sexta* larvae

Yes:

1. Measurements of feeding preferences in tobacco hornworm larvae (*Manduca sexta*) reared on three different diets
2. Can larvae of *Manduca sexta* (Arthropoda: Insecta) be induced to prefer a particular diet?

No: Effects of pollutants on sea urchin development
Yes: Influence of Cu^{++} on fertilization success and gastrulation in the sea urchin *Strongylocentrotus purpuratus*

No: Protozoan behavioral responses
Yes: Studies on the response of the protozoan *Paramecium aurelia* to shifts in light and temperature

The original titles are too vague to be compelling. Why go out of your way to give potentially interested readers an excuse to ignore your paper? Of more immediate concern in writing up laboratory reports rather than journal articles is this suggestion: why not use a title that demonstrates to your instructor that you have understood the point of the exercise? Win your reader's confidence right at the start of your report. (By the way, the title should appear on a separate page, along with your name and the date that your report is submitted.)

WRITING AN ABSTRACT

The abstract, if requested by your instructor, is placed at the beginning of your report, immediately following the title page. Yet it should be the last thing that you write since it must completely summarize the essence of your report: why the experiment was undertaken; what problem was addressed; how the problem was approached; what major results were found; what major conclusions were drawn. And it should do all this in a single paragraph. Despite its unimpressive length, a successful abstract is, therefore, notoriously difficult to write. In compact form, your abstract must present a complete and accurate summary of your work, and that summary must be fully self-contained; that is, it must make perfect sense to someone who has not read any other part of your report, as in the following example. Note that abstracts are typically written in the passive voice.

> This study was undertaken to determine the wavelengths of light that are most effective in promoting photosynthesis in the aquatic plant Elodea canadensis since some wavelengths are generally more effective than others. Rate of photosynthesis was determined at 25°C, using wavelengths of 400, 450, 500, 550, 600, 650, and 700 nm and measuring the rate of oxygen production for 1 h periods at each wavelength. Oxygen production was estimated from the rate of bubble production by the submerged plant. We tested four plants at each wavelength. The rate of oxygen production at 450 nm (approximately 2.5 ml O_2/mg wet weight of plant/h) was nearly 1.5× greater than that at any other wavelength tested, suggesting that the light of this wavelength (blue) is most readily

```
absorbed by the chlorophyll pigments. In con-
trast, light of 550 nm (green) produced no de-
tectable photosynthesis, suggesting that light of
this wavelength is reflected rather than absorbed
by the chlorophyll.
```

Note also that the sample abstract is informative. The author does not simply say that "Oxygen consumption varied with wavelength. These results are discussed in terms of the wavelengths that chlorophyll absorbs and reflects." Rather, the author provides a specific summary of the results and what they mean. Be sure that your abstract is equally informative. Clearly, this section of your report will be easiest to write if you save it for last.

PREPARING AN ACKNOWLEDGMENTS SECTION

Most biologists are aided by colleagues in various aspects of their research, and it is customary to thank those helpful people in an Acknowledgments section, the penultimate section of the report. Here is an example that might be found in a typical student report.

```
I am happy to thank Risa Cohen and Shawn Kaplan
for sharing their data with me, and Topher Gee for
late night discussions concerning the effects of
temperature on metabolic rate. Professor M. Gaudet
made me aware of the crucial Lesser and Schick
(1989) reference. Finally, I am also indebted to
Pamina Nacht for loaning me her graphics software,
and to Jerry Jarrett for teaching me how to use it.
```

As in the example above, you must include the last names of the people you are acknowledging and indicate the specific assistance received from each person named.

PREPARING THE LITERATURE CITED SECTION

In the Literature Cited section, the final section of your paper, you present the complete citations for all the factual material you refer to in the text of your report. This presentation enables the interested reader, including, perhaps, you at a later date, to obtain quickly the sources you have used in preparing your report. It provides a convenient way for the reader to obtain additional information about a particular topic, and it also provides the reader with a means of verifying what you have written as fact. It occasionally happens that a reference is used incorrectly; your interpretation or recollection of what was said in a textbook, lecture, or journal article may be wrong. By giving the source of your information, the reader can more easily recognize such errors. If the reader is your instructor, this list of references may provide an opportunity for him or her to correct any misconceptions you may have acquired. If you fail to provide the source of your information, your instructor will have more difficulty in determining where you went wrong. Proper referencing is even more crucial in scientific publications. Misstatements of fact are readily propagated in the literature by others; the Literature Cited section of a report provides the reader with the ability to verify all factual statements made, and the careful scientist consults the listed references before accepting statements made by other authors.

Listing the References

Include only those references that you have actually read and that you specifically mention in your report or paper. Unless told otherwise by your instructor, list references in alphabetical order according to the last name of the first author of each publication. If you cite several papers written by the same author, list them chronologically. If one author has published two papers in the same year, list them as, for example, C. L. Harris, 1990a, and C. L. Harris, 1990b. Each citation must include the names of all authors, the year of publication, and the full title of the paper, article, or book. In addition, when citing books, you must report the publisher, the place of publication, and the pages referred to. When citing journal articles, you must include the name of the journal, the volume number of the journal, and the page numbers of the article consulted.

There is unfortunately no single acceptable format for preparing this section of a report; formats differ from journal to journal. A few rules, however, do apply to most journals:

Spell out only the last names of authors; initials are used for first and middle names.

Latin names, including species names, are underlined to indicate italics.

Titles of journal articles are not enclosed within quotation marks.

Journal names are usually abbreviated. In particular, the word *Journal* is abbreviated as *J.,* and words ending in *-ology* are usually abbreviated as *-ol.* The *Journal of Zoology* thus becomes *J. Zool.* Do not abbreviate the names of journals whose titles are single words (for example, *Science* or *Evolution*). Acceptable abbreviations for the titles of journals can usually be found within the journals themselves.

The most important rule in preparing the Literature Cited section is to provide all the information required and to be consistent in the manner in which you present it. When preparing a paper for publication, you should religiously follow the format used by the journal to which your entry will be submitted.

The following examples should be helpful in preparing the Literature Cited section of your report. Note that the last names of all authors of a paper are included even though the names of only one or at most two authors (for example, Bayne *et al.,* 1976; Eyster and Morse, 1984) are cited in the text of the report. Also note that the format for listing books differs from the format used for citing research papers.

LISTING JOURNAL REFERENCES

Bayne, B. L. 1972. Some effects of stress in the adult on the larval development of *Mytilus edulis. Nature* (London) 237:459.

Wendt, D. E., and R. M. Woollacott. 1995. Induction of larval settlement by KCl in three species of *Bugula* (Bryozoa). *Invert. Biol.* 114: 345–351.

Woodin, S. A., S. M. Lindsay, and D. S. Wethey. 1995. Process-specific recruitment cues in marine sedimentary systems. *Biol. Bull.* 189: 49–58.

LISTING ITEMS FROM THE WORLD WIDE WEB

Brown, John. October 21, 1996. "Bugs in the News." http:/falcon.cc.ukans.edu/~jbrown/bugs/html

LISTING BOOK REFERENCES

Wessells, N. K. and J. L. Hopson. 1988. *Biology.* Random House, Inc., NY, pp. 374–379.

LISTING AN ARTICLE FROM A BOOK

Toole, B. P. 1981. Glycosaminoglycans in morphogenesis. In: *Cell Biology of Extracellular Matrix* (E. D. Hay, editor), Plenum Press, NY, pp. 259–294.

LISTING A LABORATORY MANUAL OR HANDOUT

Biology 13 Laboratory Manual. 1991. Exercise in Enzyme Kinetics, pp. 16–23. Swarthmore College, PA.

Trimmer, B. A. 1991. Principles of physiology, using insects as models. II. Excretion of organic compounds by Malphighian tubules. Biology 49 Laboratory Handout. Tufts University, Medford, MA.

A sample Literature Cited section follows, with items arranged alphabetically and chronologically. Your instructor may specify a different format for this section of your report, so check first if you are uncertain.

Literature Cited

Bayne, B. L. 1972. Some effects of stress in the adult on the larval development of Mytilus edulis. Nature (London) 237: 459.

Bayne, B. L., D. R. Livingstone, M. N. Moore, and J. Widdows. 1976. A cytochemical and biochemical index of stress in Mytilus edulis L. Mar. Poll. Bull. 7: 221-224.

Biology 13 Laboratory Manual. 1991. Exercise in Enzyme Kinetics, pp. 16-23. Swarthmore College, PA.

Eyster, L. S., and M. P. Morse. 1984. Early shell formation during molluscan embryogenesis, with new studies on the surf clam, Spisula solidissima. Amer. Zool. 24: 871-882.

Finch, C. E., and M. R. Rose. 1995. Hormones and the physiological architecture of life history evolution. Q. Rev. Biol. 70: 1-52.

Lima, G. M., and R. A. Lutz. 1990. The relationship of larval shell morphology to mode of development in marine prosobranch gastropods. J. Mar. Biol. Ass. U.K. 70: 611-637.

Toole, B. P. 1981. Glycosaminoglycans in morphogenesis. In: Cell Biology of Extracellular Matrix (E. D. Hay, editor), Plenum Press, NY, pp. 259-294.

Wessells, N. K. and J. L. Hopson. 1988. Biology. Random House, Inc., NY, pp. 374-379.

A BRIEF DISCOURSE ON STATISTICAL ANALYSIS

This brief section is no alternative to a one-semester course in biostatistics; here I will merely explain why statistics are used in Biology, what is meant by the terms *statistical analysis* and *statistical significance,* how the results of the analyses should be incorporated into the laboratory report, and how to talk about your data if you don't analyze them statistically.

What You Need to Know about Tomatoes, Coins, and Random Events

Variability is a fact of biological life: student performance on any particular examination varies among individuals; the growth rate of tomato plants varies among seedlings; the rate at which caterpillars feed on a given diet varies among individuals; the respiration rate of mice held under a given set of environmental conditions varies among individuals; the number of snails occupying a square meter of substrate varies from place to place; the amount of time a lion spends feeding varies from day to day and from lion to lion. Some of the variability we inevitably see in our data reflects unavoidable imprecision in the making of measurements. If you measure the length of a single bone 25 times to the nearest mm, for example, you will probably not end up with 25 identical measurements. But most of the variability you record in a study reflects real biological differences among the individuals in the sample population. Variability, whether it be in the responses you measure in an experiment or in the distribution of individuals in the field, is no cause for embarrassment or dismay, but it cannot be ignored in presenting your results or in interpreting them.

Suppose we plant two groups of 30 tomato seeds on day 0 of an experiment, and the individuals in group *A* receive distilled water, whereas those in group *B* receive water plus a nutrient supplement. Both groups of seedlings are held at the same temperature, are given the same volume of water daily, and receive 12 hours of light and 12 hours of darkness (12L:12D) each day for 10 days. Twenty-six of the seeds sprout under the group *A* treatment, and 23 of the seeds sprout under the group *B* treatment. At the end of 10 days, the height of each seedling is measured to the nearest 0.1 cm, and the data are recorded on the data sheet, as shown in Figure 3.16. Note that the units (cm; sample size) are clearly indicated on the data sheet, as is the nature of the measurements being recorded (height after 10 days). The number of samples taken, or of measurements made, is always represented by the symbol *N*.

The question now is this: did the mineral supplement make a difference in the height of seedlings by day 10 after planting?

If all the group *A* individuals had been 2.0 cm tall and all the group *B* individuals had been 2.4 cm tall, we would readily conclude that growth rates were increased by adding nutrients to the water. If each group *A* individual had been 2.3 cm tall and each group *B* individual had been 2.4 cm tall, we might again suggest that the nutrient supplement improved the growth rates of the seedlings. In the present case, however, there is considerable variability in the heights of the seedlings in each of the two treatments, and the difference in the average height of the two populations is not large with respect to the amount of variation found within each treatment. The heights of group *A* seedlings differ by as much as 1.0 cm (2.8 cm–1.8 cm), and

Group A seedlings: water only (height, in cm, after 10 days)
2.1 cm, 2.1, 2.0, 2.8, 2.7, 2.4, 2.3, 2.6, 2.6, 2.5, 2.1, 2.8, 2.0, 1.9, 2.8, 2.0, 2.2, 2.6, 1.8, 2.0, 2.2, 2.5, 2.4, 2.3, 2.1
Average = 2.3 cm; N = 26 measurements

Group B seedlings: water plus nutrients (height, in cm, after 10 days)
2.6 cm, 2.1, 2.0, 2.4, 2.8, 2.6, 2.2, 2.7, 2.4, 2.4, 2.3, 2.2, 2.4, 2.6, 2.4, 2.8, 2.6, 2.5, 2.6, 2.4
Average = 2.4 cm; N = 23 measurements

Figure 3.16
Data sheet with measurements recorded.

the heights of group *B* seedlings differ by as much as 0.8 cm (2.8 cm–2.0 cm), whereas the average height differences between the two groups of seedlings is only 0.1 cm (2.4 cm–2.3 cm).

The average height of the seedlings in the two populations is certainly different, but does that difference of 0.1 cm in average height reflect a real, biological effect of the nutrient supplement, or have we simply not planted enough seeds to be able to see past the variability inherent in individual growth rates? If we had planted only one seed in each group, the two seedlings might have both ended up at 2.6 cm; some seedlings reached this height in both treatment groups, as seen in Figure 3.16. On the other hand, the one seed planted in group *A* might have been the one that grew to 2.8 cm, and the one seed planted in group *B* might have been one of the seeds that grew only to 2.2 cm. Or it might have turned out the other way around, with the tallest seedling appearing in group *B*. Clearly, a sample size of one individual in each treatment would have been inadequate to conclusively evaluate our hypothesis. Perhaps 30 seeds per sample is also inadequate. If we had planted 1000 seeds, or 10,000 seeds, in each group, the differences between the two treatments might have been even less than 0.1 cm—or the differences might have been substantially greater than 0.1 cm. If only we had planted more seeds, we might have more confidence in our results. If only we had measured 100,000 individuals, or one million individuals, or. . . .

But wishful thinking has little place in Biology; we have only the data before us, and they must be considered as they stand. Is the difference between an average height of 2.4 cm for the group *A* seedlings and 2.3 cm for the group *B* seedlings a real difference? That is, is the difference statistically significant? Or have we simply conducted too little sampling to see through the variability in individual results?

As another example, suppose we have crossed plants producing yellow peas with other plants also producing yellow peas, and, from knowledge of the parentage of these two groups of pea plants, we expect their offspring to produce yellow or green peas in the ratio of 3:l. Suppose we actually count 144 offspring that produce yellow peas and 45 offspring that produce green peas so that slightly more than three times as many of the offspring produce yellow peas. Do we conclude that our expectations have been met or that they have not been met? Is a ratio of 3.2:1 close enough to our expected ratio of 3:1? Is the result (144 yellow-producing plants + 45 green-producing plants) statistically equivalent to the expected ratio?

Establishing a Null Hypothesis

Biologists use statistical tests to determine the significance of differences between sampled populations, or differences between results expected and those obtained. To begin, we must precisely define a specific issue (hypothesis) to be tested. The hypothesis to be tested is called the null hypothesis, H_0. The null hypothesis always assumes that nothing unusual has happened in the experiment or study; that is, it assumes that the treatment (addition of nutrients, for example) has no effect, or that there are no differences between the results we observed and the results we expected to observe. Examples of typical null hypotheses are

H_0: the seedlings in groups *A* and *B* do not differ in height (or the addition of nutrients does not alter growth rates of the seedlings).

H_0: the seed color of offspring does not differ from the expected ratio of 3:1.

H_0: caterpillars do not show a preference for the diet on which they have been reared.

H_0: average wing lengths do not differ among populations of house flies.

H_0: juniors did not do better than sophomores on the midterm examination.

It may seem surprising that the hypothesis to be tested is the one that anticipates no unusual effects; why bother doing the study if we begin by assuming that our treatment will be ineffective, or that there will be no differences in eye color, or that wing lengths will not differ from population to population? For one thing, the null hypothesis is chosen for testing because scientists must be cautious in drawing conclusions. Hypotheses can never be proved; they can only be discredited or supported, and the strongest statistical tests are those that discredit null hypotheses. The cautious approach in testing the effect of a new drug is therefore to assume that it will not cure the targeted ailment. The cautious approach in testing the effects of different diets on the growth rate or survival rate of a test organism is to assume that all diets will produce equivalent growth or survival—that is, that one diet is not superior to the others tested. The cautious approach in testing the effects of a pollutant is to assume that the substance is not harmful. Only if we can discredit the null hypothesis (the hypothesis of no effect) can we tentatively embrace an alternative hypothesis—for example, that a particular drug is effective, or that wing lengths do differ among populations, or that a pollutant is harmful.

Conducting the Analysis and Interpreting the Results

Once we have established our null hypothesis and collected the data for our study, statistical analysis of the data can begin. A large number of statistical tests have been developed, including the familiar Chi-Square test and the Student's *t*-test. The test that should be used to examine any particular set of data will depend on the type and amount of data collected and the nature of the null hypothesis being addressed. If you are asked to conduct a statistical analysis of your data, your laboratory instructor will undoubtedly specify the test for you. Once the appropriate test is chosen, the data are maneuvered through one or more standard, prescribed formulas to calculate the desired test statistic. This test statistic may be a Chi-Square value, a *t*-value, an F-value, or any of a variety of other values associated with different tests; in all cases, the calculated test value will be a single number, such as 0.93 or 129.8. A calculated value close to zero suggests that the data from the experiment are consistent with the null hypothesis (little deviation from the outcome expected if the null hypothesis is true). A value very different from zero indicates that the null hypothesis may be wrong since the data obtained are very different from those expected.

Returning to our seedling experiment, we wish to determine if the addition of certain nutrients alters seedling growth rates (the null hypothesis states that the nutrients have no effect). The appropriate test for this hypothesis is the *t*-test. Applying the formula provided in statistics books, the value of the *t*-statistic calculated for the data obtained in our tomato seedling experiment turns out to be −1.89. This particular value has some probability of turning up if the null hypothesis is true. Here the argument gets a bit tricky. If we repeated the experiment exactly as before, using another set of 60 seeds, we would most likely obtain a somewhat different result, and the *t*-statistic would have a different value even though the null hypothesis might still be true. If we did five identical experiments, we would probably calculate five different *t*-values from the data. In other words, a statistic may take on a broad range of values even if the null hypothesis is correct, and each of these values has some probability of turning up in any single experiment. But some values are more likely to turn up than others.

Suppose the null hypothesis, stating that the addition of nutrients does not alter the growth of tomato seedlings over the first 10 days of observation, is actually correct. If we ran our experiment (with 30 seeds planted in each of the two treatment groups) 100 times, we might actually find no measurable difference between the average heights of the seedlings in some of the experiments so that our calculated *t*-values for these data would be zero. In most of the experiments, we would probably record small differences between the average sizes of seedlings in the two populations (and, for each of these experiments, calculate a *t*-value close to zero), and in a few experiments, purely by random chance, we would probably record large differences (and calculate t-values very different from zero, either much larger or much smaller). All these results are possible if we do enough experiments even though the null hypothesis is correct, simply because the growth of seedlings is variable even under a single set of experimental conditions. The oddball result may not come up very often, but there is always some probability that it will pop up in the one experiment that we conduct.

The important point here is that the outcome of an experiment or study can vary quite a lot, whether or not the null hypothesis is actually correct. A nonbiological example may help clarify this point. In coin tossing, a fair coin should, on average, produce an equal number of heads and tails. Yet experience tells us that 10 tosses in a row will often produce slightly more of one result than the other. Every now and then, we will actually end up tossing 10 heads in a row, or 10 tails in a row, even though the coin is perfectly legitimate; neither of these results will occur very often, but each will occur eventually if we repeat the experiment enough times.

Yes, the fact of the matter is that there is considerable morphological, physiological, and behavioral variability in the real world, and that the only way to know, with certainty, that our one experiment is a true reflection of that world is to measure or count every individual in the population under consideration (for example, plant every tomato seed in the world, and measure every seedling after 10 days) or conduct an infinite number of experiments. This is not a practical solution to the problem. The next best alternative is to use statistical analysis. Statistics cannot tell us whether we have revealed THE TRUTH, but they can indicate just how convincing or just how off-the-wall our results are.

The numerical value of any calculated test statistic has some probability of turning up when the null hypothesis is true. Statisticians tell us, for example, that values of t are distributed as in Figure 3.17, and that values of Chi-Square (χ^2) are distributed as

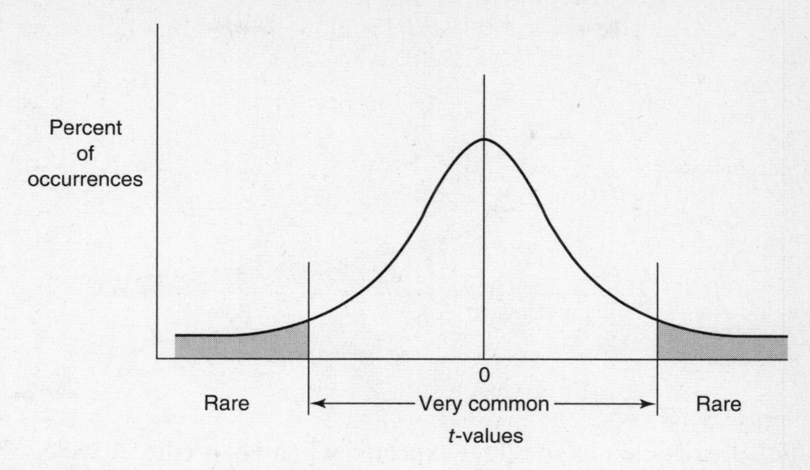

Figure 3.17
The distribution of t-values expected when the null hypothesis (H_0) is true. A wide range of values may occur, but some values will occur more commonly than others. Obtaining a common value for t causes us to accept H_0. Obtaining a rare value for t causes us to doubt the validity of H_0.

in Figure 3.18. If the null hypothesis is correct, values of each statistic will usually fall within a certain range, as indicated; these values will have the greatest probability of turning up in any individual experiment. If the t-value calculated for our experiment falls within the range indicated as "very common," we are probably safe in accepting the null hypothesis; at least we have no reason to disbelieve it. However, even if the null hypothesis is correct, very unusual values of t or of χ^2 will occasionally occur. We are, after all, randomly picking only a few seeds to plant, out of a bag that may contain many thousands of seeds; it could be just our luck to have picked only those seeds that are most unlike the average seed.

If we calculate a very unusual (very high or very low) value for t using the data from our experiment, how can we decide to reject the null hypothesis when we know there is still some small

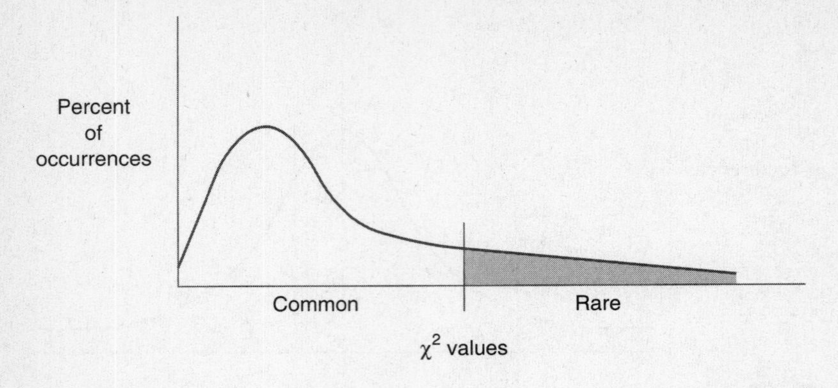

Figure 3.18
The distribution of χ^2 values expected when H_0 is true. A wide range of values may occur, but some values will occur more often than others. The rarer the value obtained, the less confidence we can have in the validity of H_0.

chance that H_0 is correct and that we have simply witnessed a very rare event? Well, we must admit that we are not omniscient and be willing to take a certain amount of risk in drawing conclusions from our data; the amount of risk taken can be specified. Typically, researchers assume that if their very unusual (that is, rarely obtained) value of t (or of some other statistic) would turn up fewer than 5 times in 100 repetitions of the same experiment when the null hypothesis is true, then this oddball value of t is a strong argument against the null hypothesis's being correct; H_0 is then tentatively rejected. That is, the large calculated value of t would be so rarely encountered if the null hypothesis is true that the null hypothesis is *probably* wrong; however, there is still the 5 in 100 chance that the null hypothesis is correct and that the researchers, through random chance, happened upon atypical results in their experiment. Tossing 10 heads in a row using a fair coin won't happen very often, but it *will* happen. Tossing 100 heads in a row is an even rarer event, but it could happen. If you conducted only one tossing experiment of 100 flips and tossed only heads, you could tentatively reject the null hypothesis that the coin is fair. But you wouldn't know with certainty that you were correct. Would you bet your car, your savings account, your little finger, or your stereo that you would get all heads if you did another round of 100 tosses? Only if the coin has two heads.

It is, I hope, becoming clear why experiments must be repeated many times before the results become convincing. Such is the challenge of doing Biology. Only when large values of a test statistic appear many times can we become fully confident that the null hypothesis deserves to be rejected. Only when low values appear many times can we become confident that the null hypothesis is most likely correct.

Summary

In performing a statistical test you first decide on a reasonable degree of risk (usually 1 or 5 percent) of incorrectly rejecting the null hypothesis, perform the study, plug the data into the appropriate formula to calculate the value of the appropriate statistic, and end up with a single number. You then look in the appropriate statistical table to see whether this number is within the range of values expected when the null hypothesis is correct. If your number lies within the expected range of values, your data support the null hypothesis. If your number lies outside the range of commonly expected values, your data do not support H_0; they support the alternative. But remember, there is always some small chance that your null hypothesis is correct and that you are making the wrong decision by rejecting H_0. Similarly, when your number lies within the range of commonly recorded values, there is always some chance your null hypothesis is actually incorrect and that you are making the wrong decision by accepting it. For this reason, we biologists never seek to *prove* any particular hypothesis; we can only accumulate data that either favor or argue against the null hypothesis.

Incorporating Statistics into Your Laboratory Report

Even if you conduct statistical analyses, be especially cautious when making statements about your data. Your data may support one hypothesis more than another, but they cannot prove that any hypothesis is true. In addition, if you conduct no statistical analyses, you cannot say that differences between groups of measurements are significant or not significant. *Significance* implies that you have subjected your data to rigorous statistical testing. It is perfectly fair to write that "seedlings treated with nutrients appeared to grow at slightly faster rates than those treated with distilled water" and refer the reader to the appropriate table or figure, but you cannot say that seedlings in one treatment grew *significantly* faster than those in the second treatment.

If you have analyzed your data using appropriate statistical procedures, the products of your heavy labor are readily and unceremoniously incorporated into the Results section of your report. Use the results of the analyses to support any major trends that you see in your data, as in the following two examples.

EXAMPLE 1

For thirty caterpillars reared on the mustard-flavored diet and subsequently given a choice of foods, the caterpillars showed a statistically significant preference for the mustard diet ($\chi^2 = 17.3$; $P < 0.05$). For thirty caterpillars reared on the quinine-flavored diet, there was no influence of previous experience on the choice of food ($\chi^2 = 0.12$; $P > 0.10$).

In this example, H_0 states that prior experience will not influence the subsequent choice of food by caterpillars; "$P < 0.05$" means that if we were to conduct the same experiment 100 times and H_0 were true, such a high value for χ^2 would be expected to occur in fewer than 5 of those 100 studies. In other words, the probability of making the mistake of rejecting H_0 when it is, in fact, true is less than 5 percent. You can therefore feel reasonably safe in rejecting H_0 in favor of the alternative: that prior experience does influence subsequent food selection for caterpillars reared on the mustard diet.

Different results were obtained, however, for the caterpillars reared on the quinine-flavored diet, $P > 0.10$ means that if the experiment were repeated 100 times and H_0 were true, you would expect to calculate such a small value of *t* in at least 10 of the 100 trials. In other words, the probability of getting this *t*-value with H_0 true is rather high; certainly the *t*-value is not unusual enough for you to mistrust H_0 and run the risk of rejecting the null hypothesis when it is, in fact, true.

EXAMPLE 2

```
Over the first 10 days of observation, growth
of seedlings receiving the nutrient supplement
was not significantly faster than the growth
of the seedlings receiving only water (t=-1.89;
P>0.10).
```

In this second example, the null hypothesis (H_0) states that the nutrient supplement does not influence plant growth; "$P > 0.10$" again means that if the experiment were repeated 100 times and H_0 were true, you would expect to calculate such a low value of *t* in more than 10 of the 100 trials. As before, you have obtained a value of *t* that would be common if H_0 were true and so have no reason to reject H_0. It is, of course, possible that H_0 is actually false and the nutrients really do promote seedling growth, and that you just happened upon an unusual set of samples that gave a misleadingly small *t*-value. If such is the case, repeating the experiment should produce different results and larger *t*-values. But with only the data before you, you cannot reject H_0.

Note that you need say little about the statistics themselves when writing your report. You would simply state, in your Materials and Methods section, that the data were analyzed by Chi-Square or some other test, and then, in the Results section, include a few test statistics to back up your interpretations of data as indicated here; the few sentences of Examples 1 and 2 tell the complete story. Statistics are used only to back up any claims you wish to make; resist the temptation to ramble on about how the statistics were calculated, how brilliant you are to have figured out which calculations to make, or how awful it was to make them.

PREPARING PAPERS FOR FORMAL PUBLICATION

Papers submitted to the editor for possible publication must conform exactly to the requirements of the specific journal you have targeted. Before beginning a manuscript, you must determine which is the most appropriate journal for your work and read carefully that journal's Instructions for Authors, typically found at the front or back of each issue or, in some cases, at the front or back of several issues each year. It also helps to study similar papers published in recent issues of the targeted journal. How are references cited in the text? How are they listed in the Literature Cited section? Does the journal permit subheadings in the Results section? If you fail to follow the relevant instructions, your paper may be returned unreviewed; at the least, you will certainly annoy the editor and reviewers.

Before mailing your manuscript to the journal's editor, go through your work one last time and be certain that every reference cited in the text is listed (and correctly so) in the Literature Cited section, and that the Literature Cited section contains no references not actually mentioned in the text. You should also indicate, in the left margin, where each figure and table is first referred to, writing something like, "Fig. 2 near here." This helps the publisher know where best to place each element.

Your manuscript should be accompanied by the correct number of copies, as specified in the Instructions to Authors section of the journal, along with a brief cover letter, which should read something like this:

> Dear Dr. Vernberg:
>
> Please consider the enclosed manuscript entitled "Influence of delayed metamorphosis on survival, growth, and reproduction of the marine polychaete Capitella sp. I" by J. A. Pechenik and T. R. Cerulli for publication in the Journal of Experimental Marine Biology and Ecology. I can send the original figures immediately upon request.
>
> Thank your for your attention.

More detailed advice about preparing professional manuscripts is given in several of the references listed in Appendix C; see, in particular, the books by R. Day and W. S. Cleveland. Note that it is usually a good idea to send the original graphs only after the manuscript has been accepted for publication; this reduces the chance of damage or loss prior to the manuscript being sent to press for printing. Photographs, however, should be included with the manuscript since the editor will need to determine whether they will reproduce well.

Checklist for the Final Draft of Your Laboratory Report

Title

____ Title gives a specific indication of what the study is about

Abstract

____ Background stated in one or two sentences
____ Clear statement of specific question addressed
____ Methods summarized in no more than three or four sentences
____ Major findings reported in no more than two or three sentences
____ Concluding sentence relates to statement of specific question addressed
____ Abstract is a single paragraph; if not, can it be rewritten as one paragraph?

Introduction

____ Clear statement of specific question or issue addressed
____ Logical argument provided as to why the question or issue was addressed
____ Every sentence leads to the statement of what was done in this study
____ All statements of fact or opinion are supported with a reference or example

Materials and Methods

____ Methods are presented in the past tense.
____ Design of study or experiment is clear and complete
____ Rationale for each step is self-evident or clearly indicated
____ Each factor mentioned is likely to have influenced the outcome of this study, and all factors likely to have influenced the outcome are mentioned
____ Includes brief description of how data were analyzed (calculations made, statistical tests used)

Continued

Results

____ Results are presented in the past tense
____ Results are presented in active terms whenever possible, for example, in terms of what organisms or enzymes did
____ All general statements are supported with reference to data
____ Major results are presented in words, but their implications are not discussed
____ The same data are not presented in both tabular and graphical form within the same report
____ Every table or graph makes an important and unique contribution to the report
____ Each figure or table has an informative caption or legend
____ Each figure or table is self-sufficient; readers can tell what question is being asked, the major aspects of how the question was addressed, and what the most important results are without reference to the rest of the paper
____ Numbers of individuals and numbers of replicates are clearly indicated in the graph, table, caption, or legend
____ The meaning of error bars on figures is clearly indicated in the caption, for example, one standard error about the mean

Discussion

____ Data are clearly related to the expectations raised in the Introduction
____ Facts are carefully distinguished from speculation
____ Unusual or unexpected findings are discussed logically, based on biology rather than apology
____ All statements of fact or opinion are supported with a reference to the literature, data, or an example
____ Discussion suggests further studies that should be conducted, additional questions that should be posed, or ways that the present study should be modified in the future

Continued

Literature Cited

____ Citations are provided for every reference cited in the report and are in the correct format
____ Section includes no references that are not cited in the report
____ Each citation includes names of all authors, title of paper, year of publication, volume number, and page numbers

Acknowledgements

____ People are mentioned by first and last names, and their specific contributions are noted

General

____ Text of report is double-spaced
____ First page shows name of author, name of lab section or instructor, and date submitted
____ All pages are numbered

4 *Writing Summaries and Critiques*

For assignments in writing summaries and critiques, you are asked to read a paper from the original scientific literature and summarize or assess that paper, usually in fewer than two double-spaced, typewritten pages. *Brief* does not, in this case, mean *easy.* In fact, producing that one- or two-page summary or critique will probably require as much mental effort as that involved in preparing an essay or term paper of from five to ten pages in length. To do well in these short assignments, you must fully understand what you have read, which usually means that you must read the paper many times, slowly and thoughtfully.

Follow the same procedures whether you are asked to write a summary or a critique; indeed, a critique begins as a summary, to which you then add your own evaluation of the paper.

To begin, read the paper once or twice without taking notes, following the advice given in Chapter 2. Fight the temptation to underline, highlight, or otherwise create the illusion that you are accomplishing something. It is often difficult to distinguish the significant from the not-so-significant points during the first reading of a scientific paper; skim the paper once for general orientation and overview. Don't try for detailed understanding in the first reading, but do jot down any unfamiliar terms or the names of unfamiliar techniques so that you can look these up in a textbook before you reread the paper.

After you have read the entire paper once, try writing down what you remember about the paper, what you don't understand about what you read, and any other questions that come to mind as you write. This will help to focus your attention on some of the major points for a second reading. It often helps to consult a textbook about the general biology of the organisms being studied before returning to the paper.

During the next, more careful reading of the paper, pay special attention to the Materials and Methods and the Results sections; the essence of any scientific paper is contained here. The results obtained in a study depend on how the study was conducted. Were samples taken only at one particular time of year? Was the study replicated? How many individuals were examined? What techniques were used? In an experiment, what variables (for example, photoperiod, temperature, salinity, or food supply) were held constant? Were proper controls provided for each experiment? Which factors might affect the outcome of the study?

As you begin to study the Results section, scrutinize every graph, table, and illustration, developing your own interpretations of the data before rereading the author's verbal presentation, as discussed in Chapter 2. We are readily influenced by the opinions of others, especially when those opinions are well written. Keep an open mind when reading the author's words, but try to form your own opinions about the data first; you may see something that the author did not.

WRITING THE FIRST DRAFT

You will know that you are ready to write your first draft of the assignment when you can distill the essence of the paper into a single, intoxicating summary sentence, or, at most, two summary sentences, as discussed in Chapter 2. These sentences should include *all* the key points, present an *accurate* summary of the study, be *fully comprehensible* to someone who has never read the original paper, and be in your own words. As a general rule, do not begin to write your review until you can write such an abbreviated summary; this exercise will help you discriminate between the essential points of the paper and the extra, complementary details. Several examples of good summary sentences are given later.

If you cannot write a satisfactory one- or two-sentence summary, reread the article; you'll get it eventually. Once your summary sentence is committed to paper, ask yourself these questions:

1. Why was the study undertaken? To answer this, draw especially from information given in the Introduction and Discussion sections of the paper you have read.
2. What specific questions were addressed? Summarize each question in a single sentence.
3. How were these questions addressed? What specific approaches were taken to address each question on your list?
4. What were the major findings of the study?
5. What questions remain unanswered by the study? These may be questions addressed by the study but not answered conclusively, or they may be new questions arising from the findings of the study under consideration.

WRITING THE SUMMARY

When you can answer these questions without referring to the paper you have read, you can begin to write.

At the top of the page—below your name, the course designation, and the date—give the complete citation for the paper being discussed, beginning at the left-hand margin: names of all authors, year of publication, title of the paper, title of the journal in which the paper was published, and volume and page numbers of the article. On a new line, indent five spaces and begin your summary with a few sentences of background information. Your introductory sentences must lead up to a statement of the specific questions the researchers set out to address. Next, tell (1) what approaches were used to investigate each question and (2) what major results were obtained. Be sure to state, as succinctly as possible, exactly what was learned from the study.

To cover so much ground within the limits of one typewritten page is no small feat, but it can be done if you first make certain that you fully understand what you have read. Consider the following example of a brief, successful summary. Before writing the summary, the student condensed the paper into these two sentences.

> The tolerance of a Norwegian beetle (Phyllodecta laticollis) to freezing temperatures varied seasonally, in association with changes in the blood concentration of glycerol, amino acids, and total dissolved solute. However, the concentration of nucleating agents in the blood did not vary seasonally.

Note that the two-sentence distillation contains considerable detail despite its brevity, implying impressive mastery of the paper's contents; it is complete, accurate, and self-sufficient. When you can write such sentences, pat yourself on the back and proceed; the hardest work is over.

SAMPLE STUDENT SUMMARY

MINNIE LEGGS
Bio 101
April 12, 1996

Van der Laak, S. 1982. Physiological adaptations to low temperature in freezing-tolerant Phyl-

lodecta laticollis beetles. *Comp. Biochem. Physiol.* 73A: 613-620.

Adult beetles (*Phyllodecta laticollis*), found in Norway, are exposed to sub-zero (°C) temperatures in the field throughout the year. In general, organisms that tolerate freezing conditions either produce extracellular nucleating agents that trigger ice formation outside the cells rather than within them or they produce biological antifreezes, such as glycerol, that lower the freezing point of the blood and tissues to below that of the environment, thereby preventing ice formation. This study was undertaken to document the tolerance of *P. laticollis* to below-freezing temperatures and to account for seasonal shifts in the temperature tolerance of these beetles.

Beetles were collected throughout the year and frozen to temperatures as low as −50°C; post-thaw survivorship was then determined. Determinations were also made of the concentrations of solutes in the blood (that is, blood osmotic concentration), total water content, amino acid and glycerol concentrations in the blood, presence of nucleating agents in the blood, and the temperature to which blood could be super-cooled before freezing would occur.

The temperature tolerance of *P. laticollis* varied from about −9°C in summer to about −42°C in winter; this shift in freezing tolerance was paralleled by a dramatic winter increase in glycerol concentration and in total blood osmotic concentration. Amino acid concentration also increased in winter, but the contribution to blood osmolarity was small compared to that of glycerol. Nucleating agents were present in the blood year-round, ensuring that ice formation will occur extracellularly rather than intracellularly, even in summer.

For beetles collected in midwinter and early spring, blood glycerol concentrations could be artificially reduced by warming beetles to 23°C (room temperature) for about 24-150 h. When glycerol concentrations of spring and winter beetles were reduced to identical levels by warming, the spring beetles tolerated freezing better than the winter beetles; these differences in tolerance could not be explained by differences in amino acid concentrations. This result indicates that some other factors, as yet unknown, are also involved in determining the freezing tolerance of these beetles.

Analysis of Student Summary

The student has, within one typed page, successfully distilled a seven-page technical report to its scientific essence. Note that the student used the first three sentences to introduce the topic and then summarized the purpose of the research in one sentence. The next short paragraph summarizes the experimental approach taken, and the main findings of the study are then stated. No superfluous information is given; the author of this assignment provided only enough detail to make the summary comprehensible. The product glistens with understanding. Rereading the student's two-sentence encapsulation of the paper (p. 144), you can see that the student was indeed ready to write the report.

As a challenge to yourself, try writing a one-paragraph summary of the study example, cutting the length of the original summary by about 75 percent. Summary is the ultimate test of understanding.

WRITING THE CRITIQUE

A critique is much like a summary, except that you get to add your own assessment of the paper you have read. This does not mean you should set out to tear the paper to shreds; a critical review is a thoughtful summary and analysis, not an exercise in character assassination. Almost every piece of biological research has shortcomings, most of which become obvious only in hindsight. Yet every piece of research contributes some information, even when the original goals of the study are not attained. Emphasize the positive—focus on what was learned from the study. Although you should not dwell on the limitations of the study, you should point out these limitations toward the end of your critique. Were the conclusions reached by the authors out of line with the data presented? Do the authors generalize far beyond the populations or species studied? Which questions remain unanswered? How might these questions be addressed? How might the study be improved or expanded in the future? Keep this in mind as you write: you wish to demonstrate to your instructor (and to yourself) that you understand what you have read. Do not comment on whether or not you enjoyed the paper, or found it to be well written; stick to the science unless told otherwise.

THE CRITIQUE

Before writing the critique, the student produced this one-sentence summary of the paper.

> The egg capsules of the marine snails *Nucella lamellosa* and *N. lima* protect developing embryos against low-salinity stress, even though the solute concentration within the capsules falls to near that of the surrounding water within about one h.

Again, note that this one-sentence summary satisfies the criterion of self-sufficiency: it can be fully understood without reference to the paper it summarizes. The critique follows.

SAUL TEE
Bio 101

Kînehcép, N. A. 1982. Ability of some gastropod egg capsules to protect against low-salinity stress. *J. Exp. Marine Biol. Ecol.* 63: 195-208.

The fertilized eggs of marine snails are often enclosed in complex, leathery egg capsules with 30 or more embryos being confined within each capsule. The embryos develop for one or more weeks before leaving the capsules. The egg capsules of intertidal species potentially expose the developing embryos to thermal stress, osmotic stress, and desiccation stress. This paper describes the ability of such egg capsules to protect developing embryos from low-salinity stress, such as might be experienced at low tide during a rainstorm.

Two snail species were studied: Nucella lamellosa and N. lima. Embryos were exposed, at 10-12°C, either to full-strength seawater (control conditions) or to 10-12% seawater solutions (seawater diluted with distilled water). The ability of egg capsules to protect the enclosed embryos from low-salinity stress was assessed by placing intact egg capsules into the test solutions for up to 9 h, returning the capsules to full-strength seawater, and comparing subsequent embryonic mortality with that shown by embryos removed from capsules and exposed to the low-salinity stress directly.

Encapsulated embryos exposed to the low salinities suffered less than 2% mortality, even after low-salinity exposures of 9 h duration. In contrast, embryos exposed directly to the same test conditions for as little as 5 h suffered 100% mortality. All embryos survived exposure to control conditions for the full 9 h, showing that removal from the capsules was not the stress killing the embryos in the other treatments. Sampling capsular fluid at various times after capsules were transferred to the diluted seawater, Kînehcép found that the concentration of solutes within capsules fell to near that of the surrounding water within about one h after transfer.

This study clearly demonstrates the protective value of the egg capsules of two snail species faced with low-salinity stress. However, Kînehcép was unable to explain how egg capsules of these two species protect the enclosed embryos, since the capsules did not prevent decreases in the solute concentration of the capsular fluid. Although Kînehcép plotted the rate at which the solute concentration falls within the capsules (his Fig. 1), he sampled only at 0, 60, and 90 minutes after the capsules were transferred to water of reduced salinity. I think he should have sampled at frequent intervals during the first 60 min to discover how rapidly the solute concentration of the capsule fluid falls. As Kînehcép himself suggests, perhaps the embryos are less stressed if the concentration inside the capsule falls slowly. These experiments were all performed at a single temperature even though encapsulated embryos are likely to experience fluctuation in both temperature and salinity as the tide rises and falls during the day; the study should be repeated using a range of temperatures likely to be experienced in the field. In addition, I suggest repeating these experiments using deep-water species whose egg capsules are never exposed to salinity fluctuations of the magnitude used in this study.

Analysis of Student Critique

As before, this student begins with just enough introductory information to make the point of the study clear and ends the first paragraph with a succinct statement of the researcher's goal. The methods and results of the study are then briefly reviewed, as in a summary. Whereas a summary would probably end at this point, the critique continues with thought-provoking assessments by the student. Note that the student was careful to distinguish his or her thoughts from those of the paper's author (see pp. 28–30, on plagiarism).

CONCLUDING THOUGHTS

Clearly, successfully completing either type of assignment is no trivial matter. But preparing good summaries and critiques is an excellent way to push yourself toward true understanding of what you read, and of the nature of scientific inquiry.

5
Writing Essays and Term Papers

A term paper is really just a long essay, its greater length reflecting more extensive treatment of a broader issue. Both assignments ask you to present critical evaluations of what you have read. In preparing an essay, you synthesize information, explore relationships, analyze, compare, contrast, evaluate, and organize your own arguments clearly, logically, and persuasively, gradually leading up to an assessment of your own. A good term paper or short essay is a creative work; you must interpret thoughtfully what you have read and come up with something that goes beyond what is presented in any single article or book consulted.

Essays and term papers are based mostly on readings from the primary scientific literature—that is, the original research papers published in such scientific journals as *Biological Bulletin, Cell, Developmental Biology, Ecology,* and *Journal of Comparative Biochemistry and Physiology.* Textbooks and review articles (such as those in *Scientific American* or *Quarterly Review of Biology*) compose the secondary literature. The secondary literature gives someone else's interpretation and evaluation of the primary literature. In preparing an essay or term paper, you will go through the same processes that the writers of textbooks and review articles go through in presenting and discussing the primary scientific literature.

WHY BOTHER?

Every time you are asked to write an essay or term paper, your instructor is committing himself or herself to many hours of reading and grading. There must be a good reason to require such assignments; most instructors are not masochists.

In fact, writing essays or term papers benefits you in several important ways. For one thing, you end up teaching yourself something relevant to the course you are taking. The ability to self-teach is essential for success in graduate programs and academic careers, and is a skill worth cultivating for success in almost any profession. Additionally, you gain experience in reading the primary scientific literature. Textbooks and many lectures present you with facts and interpretations. By reading the papers upon which these facts and interpretations are based, you come face-to-face with the sorts of data, and interpretations of data, that the so-called facts of Biology are based on, and you gain insight into the true nature of scientific inquiry. The data collected in an experiment are always real; interpretations, however, are always subject to change. Preparing thoughtful essays and term papers will help you move away from the unscientific, blind acceptance of stated facts toward the scientific, critical evaluation of data. These assignments are also superb exercises in the logical organization, effective presentation, and discussion of information, skills that can only ease your career progress in the future. How fortunate you are that your instructor cares enough about your future to give such assignments!

There is one last reason that instructors often ask their students to prepare essays. One can simply summarize a dozen papers in succession without understanding the contents of any of them. I call this the book report format, in which the writer merely presents facts uncritically: this happened, that happened; the authors suggested this; the authors found that. By writing an essay rather than a book report, you can show your instructor that you really understand what you have read, that you have really learned something rather than simply memorized or mimicked the information presented to you.

GETTING STARTED

You must first decide on a general subject of interest. Often your instructor will suggest topics that have been successfully exploited by former students. Use these suggestions as guides, but do not feel compelled to select one of these topics unless so instructed. Be sure to choose or develop a subject that interests you. It is much easier to write successfully about something of interest than about something that bores you.

All you need for getting started is a general subject, not a specific topic. Stay flexible. As you research your selected subject, you will usually find that you must narrow your focus to a particular topic because you encounter an unmanageable number of references pertinent to your original idea. You cannot, for instance, write about the entire field of primate behavior because the field has many different facets, each associated with a large and growing literature. In such a case, you will find a smaller topic, such as the social significance of primate grooming behavior, to be more appropriate; as you continue your literature search, you may even find it necessary to restrict your attention to a few primate species.

Alternatively, you may find that the topic originally selected is too narrow and that you cannot find enough information on which to write a substantial paper. You must then broaden your topic, or switch topics entirely, so that you will end up with something to discuss. Don't be afraid to discard a topic on which you can find too little information.

Choose a topic you can understand fully. You can't possibly write clearly and convincingly on something beyond your grasp. Don't set out to impress your instructor with complexity; instead, dazzle your instructor with clarity and understanding. Simple topics often make the best ones for essays and papers.

RESEARCHING YOUR TOPIC

As discussed in Chapter 2, begin by carefully reading the appropriate section of your textbook to get an overview of the general subject of which your topic is a part, and then consult at least one more specialized book before tackling the primary literature. Now you are ready to locate and read research reports on your topic, following the advice presented in Chapter 2. The goal is to select a number of interrelated papers and to read these with considerable care and patience, not to simply accumulate a huge number of references that then receive cursory attention. Continually ask yourself while taking notes, "Why am I writing this down? What is especially interesting about this particular information? Can I see any relationship between this information and what I have already written or learned?"

Be sure you understand thoroughly what you have read. One of the best ways to self-assess your understanding is to summarize the material as you read along, paragraph by paragraph, section by section. When you have completed the entire paper, try writing a one-paragraph summary of what you have read, and then a one- or two-sentence summary. You cannot evaluate or synthesize information until you can first write a clear, accurate, and specific summary of that information in your own words, as discussed more fully in Chapters 2 and 4: summary is a necessary prelude to synthesis and critical evaluation.

WRITING THE PAPER

Begin by reading all of your notes, preferably without pen or pencil in hand. Having completed a reading of your notes to get an overview of what you have accomplished, reread them, this time with the intention of sorting your ideas into categories. Notes taken on index cards are particularly easy to sort, provided that you have not written many different ideas on a single card; one idea per card is a good rule to follow. To arrange notes written on full-sized sheets of paper, some people suggest annotating the notes with pens of different colors or using a variety of symbols, with each color or symbol representing a particular aspect of the topic. Still other people simply use scissors to snip out sections of the notes, and then group the resulting scraps of paper into piles of related ideas. You should experiment to find a system that works well for you.

At this point, you must eliminate those notes that are irrelevant to the specific topic you have finally decided to write about. No matter how interesting a fact or idea is, it has no place in your paper unless it clearly relates to the rest of the paper and therefore helps you develop your argument. Some of the notes you took early on in your exploration of the literature are especially likely to be irrelevant to your essay since these notes were taken before you had developed a firm focus. Put these irrelevant notes in a safe place for later use; don't let them coax their way into your paper.

You must next decide how best to arrange your categorized notes so that your essay or term paper progresses toward some conclusion. Again, ask yourself whether a particular section of your notes seems interesting to you, and why it does, and look for connections among the various items as you sort.

The direction your paper will take should be clearly and specifically indicated in the opening paragraph, as in the following example written by student *A*:

> Most lamellibranch bivalves are sedentary, living either in soft-substrate burrows (e.g., soft-shell clams, Mya arenaria) or attached to hard substrate (e.g., the blue mussel Mytilus edulis) (Barnes, 1980). However, individuals of a few bivalve species live on the surface of substrates, unattached, and are capable of locomoting through the water. One such species is the scallop Pecten

> maximus. In this essay, I will explore the morphological and physiological adaptations that make swimming possible in P. maximus, and will consider some of the evolutionary pressures that might have selected for these adaptations.

The nature of the problem being addressed is clearly indicated in this first paragraph, and student *A* tells us clearly why the problem is of interest: (1) the typical bivalve doesn't move and certainly doesn't swim, (2) a few bivalves can swim, (3) so what is there about these exceptional species that enables them to do what other species can't? (4) and why might this swimming ability have evolved? Note that use of the pronoun *I* is now perfectly acceptable in scientific writing.

In contrast to the previous example, consider the following weaker (although not horrible) first paragraph written by student *B* on the same subject.

> Most bivalved mollusks either burrow into, or attach themselves to, a substrate. In a few species, however, the individuals lie on the substrate unattached and are able to swim by expelling water from their mantle cavities. One such lamellibranch is the scallop Pecten maximus. The feature that allows bivalves like P. maximus to swim is a special formation of the shell valves on their dorsal sides. This formation and its function will be described.

In this example, the second sentence weakens the opening paragraph considerably by prematurely referring to the mechanism of swimming. The main function of the sentence should be to emphasize that some species are not sedentary; the reader, not yet in a position to understand the mechanism of swimming, becomes a bit baffled. The next-to-last sentence of the paragraph ("The feature that allows . . .") also hinders the flow of the argument. This sentence summarizes the essay before it has even been launched, and again, the reader is not yet in a position to appreciate the information presented; what is this "special formation," and how does it have anything to do with swimming? The first paragraph of a paper should be an introduction, not a summary.

The last sentence of student *B*'s paragraph does clearly state the objective of the paper, but the reader must ask, "Toward what end?" The author has set the stage for a book report, not an essay. Reread the paragraph written by student *A,* and notice how the same information has been used much more effectively, introducing a thoughtful essay rather than a book report. Student *A*'s first paragraph was written with a clear sense of purpose, with each sentence carrying the reader forward to the final statement of intent. You might guess (correctly, as it turns out) from reading student *B*'s first paragraph that the rest of the paper was somewhat unfocused and rambling. In contrast, student *A*'s first paragraph clearly signals that what follows will be well focused and tightly organized. Get your papers off to an equally strong start.

Another example of a typical, but not especially effective, first paragraph will be helpful.

> The crustaceans have an exceptional capability for changing the intensity and pattern of their coloring (Russell-Hunter, 1979). Many species seem able to change their color at will. The cells responsible for producing the characteristic color changes of crustaceans are the chromatophores. The function of these cells will be discussed in this essay.

What is wrong with this introductory paragraph? The author is certainly off to a strong start with the first sentence. The second sentence, however, begins by repeating information already given in the first sentence (crustaceans can change color), and ends by saying nothing at all (what does "at will" mean for a crustacean?). The last sentence sets up a book report even though the author calls it an essay. Why will the function of these cells be discussed? More to the point, why should the reader be interested in such a discussion? The reader will be more readily drawn into your intellectual net if you indicate not only where you are heading but also why you are undertaking the journey.

The first paragraph of your paper must state clearly what you are setting out to accomplish and why. Every paragraph that follows the first paragraph should advance your argument clearly and logically toward the stated goal.

State your case, and build it carefully. Use your information and ideas to build an argument, to develop a point, to synthesize. Avoid the tendency to simply summarize papers one by one: the authors did this, then they did that, and then they suggested the following explanation. Instead, set out to compare, to contrast, to illustrate, to discuss. As in all other scientific writing, always back up your statements with supporting documentation; this documentation may be an example drawn from the literature you have read or just a reference (author and date of publication) to a paper or group of papers that support your statement, as in the following examples:

> There is no evidence for mechanical ventilation in freshwater pulmonates, so it is presumed that gases exchange solely by diffusion (Ghiretti and Ghiretti-Magaldi, 1975).

> Schistosomiasis is one of the most serious parasitic diseases of mankind, afflicting hundreds of millions of people and causing hundreds of millions of dollars in economic losses yearly through livestock infestations (Noble and Noble, 1982).

> The ability of an organism to recognize "self" from "non-self" is found in both vertebrates and invertebrates. Even the most primitive invertebrates

show some form of this immune response. For example, Wilson (1907) found that disassociated cells from two different sponge species would regroup according to species; cells of one species never reaggregated with those of the second species.

In referring to experiments, don't simply state that a particular experiment supports some particular hypothesis; describe the relevant parts of the experiment, and explain how the results relate to the hypothesis under question, as demonstrated in the following two examples:

EXAMPLE 1

Foreign organisms or particles that are too large to be ingested by a single leukocyte are often isolated by encapsulation, with the encapsulation response demonstrating clear species-specificity. For example, Cheng and Galloway (1970) inserted pieces of tissue taken from several species into an incision made in the body wall of the gastropod <u>Helisoma duryi</u>. Tissue transplanted from other species was completely encapsulated within 48 hours of the transplant. Tissue obtained from individuals of the same species as the host was also encapsulated, but encapsulation was not completed for at least 192 hours.

EXAMPLE 2

Above a certain temperature, further temperature increases often have a depressing effect on larval growth rates (Kingston, 1974; Leighton, 1974). This break point can be very sharply defined. For instance, larvae of the bivalve Cardium glaucum were healthy and grew rapidly at 31°C, grew abnormally and less rapidly at 32-33°C, and grew hardly at all at 34°C (Kingston, 1974).

In all of your writing, avoid quotations unless they are absolutely necessary; use your own words whenever possible. At the end of your essay, summarize the problem addressed and the major points you have made, as in the following example:

> Clearly, the basic molluscan plan for respiration that had been successfully adapted to terrestrial life in one group of gastropods, the terrestrial pulmonates, has been successfully readapted to life in water by the freshwater pulmonates. Having lost the typical molluscan gills during the evolutionary transition from salt water to land, the freshwater pulmonates have evolved new respiratory mechanisms involving either the storage of an air supply (using the mantle cavity) or a means of extracting oxygen while under water, using a gas bubble or direct cutaneous respiration. Further studies are required to fully understand how the gas bubble functions in pulmonate respiration.

Never introduce any new information in your summary paragraph.

CITING SOURCES

This topic has been covered briefly in Chapter 3 (pp. 95–98). Cite only those sources you have actually read and would feel confident discussing with your instructor. Unless told otherwise, do not cite these sources with footnotes. Instead, cite them directly in the text by author and date of publication. For example:

> Kim (1976) demonstrated that magnetic fields established by direct current can alter the rates of enzyme-mediated reactions in cell-free systems. Similarly, magnetic fields established by alternating current can affect the activity of certain liver enzymes (Yashina, 1974) and mitochondrial enzymes (Kholodau, 1973).

Try to make the relevance of the cited reference clear to the reader. For example, rather than writing,

> Temperature tolerances have been determined for gastropods, bivalves, annelids, and insects (Gable, 1988; Booth, 1989; Booth and Gable, 1993).

it would be clearer to write,

> Temperature tolerances have been determined for gastropods (Gable, 1988), bivalves and annelids (Booth, 1989), and insects (Booth and Gable, 1993).

As always, if you write as though explaining something to yourself, your classmates, or your parents, or if you write so that the paper will be useful to you in the future, you will generally come out ahead.

At the end of your paper, include a section entitled Literature Cited, listing all references you have referred to in your paper. Do not include any references you have not actually read. Each reference listed must give author(s), date of publication, title of article, title of journal, and volume and page numbers. If the reference is a book, the citation must include the publisher, place of publication, and total number of pages in the book, or the page numbers pertinent to the citation. Your instructor may specify a particular format for preparing this section of your paper. For additional information about preparing the Literature Cited section, see Chapter 3 (pages 120–123).

CREATING A TITLE

By the time you have finished writing, you should be ready to title your creation. Give the essay or term paper a title that is appropriate and interesting, one that conveys significant information about the specific topic of your paper (see also pages 116–117).

No: Factors controlling sex determination in turtles
Yes: The roles of nest site selection and temperature in determining sex ratio in loggerhead sea turtles

No: Biochemical changes during hibernation
Yes: Adaptations to environmental stress: the biochemical basis for depressed metabolic rate in hibernating mammals

No: Molluscan defense mechanisms
Yes: Behavioral and chemical defense mechanisms of gastropods and bivalves

No: Asexual reproduction: is it the way to go?
Yes: Sexual and asexual reproduction in cnidarians: an evaluation of advantages and disadvantages associated with two different modes of reproduction

REVISING

Once you have a working draft of your paper, you must revise it, clarifying your presentation, removing ambiguity, eliminating excess words, and improving the logic and flow of ideas. You may also have to edit for grammar and spelling. These topics are considered in detail in Chapter 10.

6
Writing Research Proposals

Research proposals are commonly assigned in advanced Biology courses in place of the more standard "term paper"; the two assignments have much in common, and you should read, or reread, Chapter 5 before proceeding with this chapter. Research proposals, essays, and term papers all involve critical review and synthesis of the primary literature—that is, papers presenting detailed, original results of research rather than articles and books presenting only summaries and interpretations of that research. In addition, however, a research proposal includes a written argument in which you propose to go beyond what you have read; you propose to do a piece of research yourself and seek to convince the reader that what you propose to do should be done and can be done (Figure 6.1), and should in fact be done exactly how you propose to do it. Research proposals are perhaps the best vehicle for developing your reasoning and writing skills in Biology. This assignment, more than any other, gives you a chance to be creative and to become a genuine participant in the process of biological investigation. Writing a good research proposal is no trivial feat, and the sense of accomplishment you feel once you are finished is indescribably nice.

Research proposals have two major parts: a review of the relevant scientific literature and a description of the proposed research. In the first part, you review the primary literature on a particular topic, but you do so with a particular goal in mind: you wish to lead your reader to the inescapable conclusion that the question you pro-

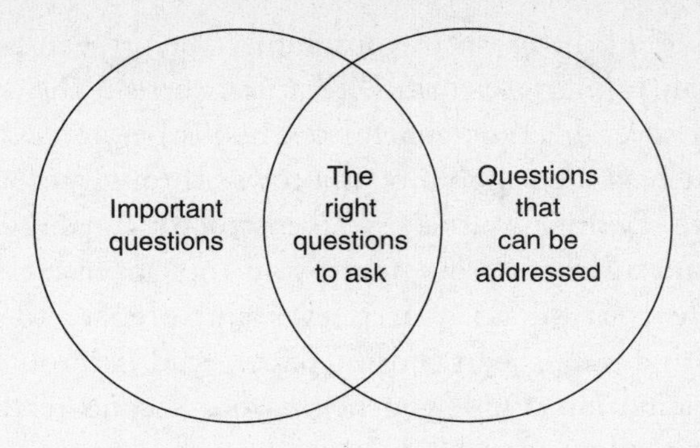

Figure 6.1
The trick of developing a valid research question. Many questions are easy to answer but are meaningless or too trivial to be worth asking, whereas other questions are important but unapproachable by existing methods.

pose to address follows logically from the research that has gone before. Writing a research proposal rather than a term paper thus helps you avoid falling into the book report trap; once you develop a research question to ask, you should have an easier time focusing your literature review on the development of a single, clearly articulated theme. Developing that theme will take some time and thought, but your writing will then have a clear direction.

In addition to providing you with a convenient vehicle for exploring and digesting the primary scientific literature and for focusing your discussion of that literature, you may find that the research you propose to do can actually be done—and can be done by you. Your proposal could turn out to be the basis for your own summer research project, senior thesis, or even master's or Ph.D. thesis.

RESEARCHING YOUR TOPIC

Proceed as you would for researching a term paper or essay (see Chapter 5). For this assignment especially, it will be important to have a firm grasp of your subject before you plunge into the original, primary scientific literature, so read the appropriate sections of several recent general textbooks before you look elsewhere. The next step should be to browse through recent issues of appropriate scientific journals; your instructor can suggest several that are particularly appropriate to your topic of interest.

Before you roll up your sleeves and prepare to wrestle in earnest with a published scientific paper, read it through once for general orientation. Once you begin your second reading of the paper, don't allow yourself to skip over any sentences or paragraphs you don't understand. Keep a relevant textbook by your side as you read the primary literature so that you can look up unfamiliar facts and terminology.

I mentioned previously that the results of a study depend largely on the way the study was conducted. We have also seen that although the results of a study are real, the interpretation of those results is always subject to change. The Materials and Methods section and the Results section of research papers must therefore be read with particular care and attention, as discussed in Chapter 2. Scrutinize every table and graph until you can reach your own tentative conclusions about the results of the study before allowing yourself to be swayed by the author's interpretations. Read with a questioning, critical eye, as discussed in Chapter 2.

As you carefully read each paper, pay special attention to the following:

1. What specific question is being asked?
2. How does the design of the study address the question posed?
3. What are the controls for each experiment? Are they appropriate and adequate?
4. How convincing are the results? Are any of the results surprising?
5. What contribution does this study make toward answering the original question?

6. What aspects of the original question remain unanswered?

Reread the paper until you can answer each of these questions. Then ask yourself the following additional question:

7. What might be a next logical question to ask, and how might this question be addressed?

Continue your library research using the references listed at the end of the recent papers you are reading, and perhaps by consulting *Biological Abstracts* or one of the other indexing services discussed in Chapter 2. One particularly convenient thing about preparing a research proposal is that it's relatively easy to tell when your library work is finished; it's finished when you know what your proposed research question will be and when you know exactly why you are asking that question.

WRITING THE PROPOSAL

Divide your paper into three main portions: Introduction, Background, and Proposed Research.

Introduction

Give a brief overview of the research being considered, and indicate the nature of the specific question you will pursue, as in the following example*:

> Studies have shown that such endurance exercises as running and swimming can affect the reproductive physiology of women athletes. Female runners (Dale et al., 1979; Wakal et al., 1982), swimmers (Frisch et al., 1981), and ballet dancers

*Modified from a student paper written by A. Lord.

(Warren, 1980) menstruate infrequently (oligomenorrhea) in comparison with nonathletic women of comparable age, or not at all (amenorrhea). The degree of menstrual abnormality varies directly with the intensity of the exercise. For example, Malina _et al._ (1978) have shown that menstrual irregularity is more common, and more severe, among tennis players than among golfers.

The physiological mechanism through which strenuous activity disrupts the normal menstrual cycle is not yet clear; inadequate fat levels (Frisch _et al._, 1981), altered hormonal balance (Sutton _et al._, 1973), and physiological predisposition (Wakat _et al._, 1982) have each been implicated.

In the proposed research, I will study 175 women weight lifters in an attempt to determine the relative importance of fat levels, hormone levels, and physiological predisposition in promoting oligomenorrhea and amenorrhea.

Notice that the author of this proposal has not used the Introduction to discuss the question being addressed or to describe how the study will be done. The Introduction provides only (1) general background to help the reader understand why the topic is of interest, and (2) a brief but clear statement of the specific research topic that will be addressed. It helps to write the last sentence of your Introduction first, stating the specific question to be addressed; then write the rest of your Introduction, giving just enough information for the reader to understand why anyone would want to ask such a question. Limit your introduction to two or three paragraphs. A detailed discussion of prior research belongs in the Background section of the proposal, and a detailed description of the proposed study belongs in the Proposed Research section of the proposal.

Notice also that every factual statement (for example, "Female runners . . . menstruate infrequently") is supported by a reference to one or more papers from the primary literature. These references enable the reader to obtain, painlessly, additional information on particular aspects of the subject and to verify the accuracy of statements made in the proposal. Backing up statements with references also protects the author of the proposal by documenting the source of information; if the author of your source is mistaken, why should you take the blame?

Background

In this section, you demonstrate your complete mastery of the relevant literature. Discuss this literature in detail, leading up to the specific objective of your proposed research. This section of your proposal follows the format of a good term paper or essay, as already described in Chapter 5 (pp. 155–161). In a proposal, however, the Background section will end with a brief statement of what is now known and what is not yet known about the research topic under consideration, and a clear, specific description of the research question(s) you propose to investigate. Here are two examples. The author of Example 1 has already spent two and a half pages of the Background section describing documented effects of organic pollutants on adults and developmental stages of various marine vertebrates and invertebrates.

EXAMPLE 1

Thus many fish, echinoderm, polychaete, mollusc, and crustacean species are highly sensitive to a variety of fuel oil hydrocarbon pollutants, and the early stages of development are especially susceptible. However, many of these species begin their lives within potentially protective extra-embryonic egg membranes, jelly masses, or egg capsules (Anderson et al., 1977; Eldridge et al., 1977; Kînehcép, 1979). The ability of these structures to protect developing embryos against water-soluble toxic hydrocarbons has apparently never been assessed.

The egg capsules of marine snails are particularly complex, both structurally and chemically (Fretter, 1941; Bayne, 1968; Hunt, 1971). Such capsules are typically several mm to several cm in height, and the capsule walls are commonly 50-100 μm* thick (Hancock, 1956; Tamarin and Carriker, 1968). Depending on the species, embryos may spend from several days to many weeks developing within these egg capsules before emerging as free-swimming larvae or crawling juveniles (Thorson, 1946).

Little is known about the tolerance of encapsulated embryos to environmental stress, or about the permeability of the capsule walls to water and solutes. Kînehcép (1982, 1983) has found that the egg capsules of several shallow-water marine snails (Ilyanassa obsoleta, Nucella lamellosa, and N. lapillus) are permeable to both salts and water, but are far less permeable to the small organic molecule glucose. Capsules of at least these species are thus likely to protect embryos from exposure to many fuel oil components.

In the proposed study, I will (1) document the tolerance of early embryos of N. lamellosa and N. lapillus, both within capsules and removed from

*μm = microns (10^{-6} meters).

capsules, to the water soluble fraction of Number 2 fuel oil; (2) determine the general permeability characteristics of the capsules of these two gastropod species to see which classes of toxic substances might be unable to penetrate the capsule wall; and (3) use radioisotopes to directly measure the permeability of the capsules to several major components of fuel oil.

The shorter Example 2 concerns the hormonal control of reproductive activity in sea stars. The author of this proposal has already spent three pages of the Background section discussing experiments demonstrating that (1) gamete release (spawning) is under hormonal control; (2) the response to the hormone varies seasonally; and (3) the variation in response seems to reflect changing concentrations of an inhibitory hormone called shedhibin.

EXAMPLE 2

The present evidence suggests, therefore, that the influence of the excitatory hormone is regulated by seasonal fluctuations in the secretion of shedhibin, although seasonal changes in the concentrations of this inhibitory hormone have not yet been documented.

In the proposed research, I will:

(1) identify the time of year during which gamete release is inhibited in mature sea stars (<u>Asterias forbesi</u>);

(2) develop a monoclonal antibody to the inhibitory substance shedhibin;

(3) and use immunofluorescent techniques to quantify the amount of shedhibin produced and secreted at different stages of the reproductive cycle of A. forbesi.

Proposed Research

This portion of your proposal has two interrelated parts: (1) what specific question(s) will you ask? and (2) how will you address each of these questions? Different instructors will put different amounts of stress on these two parts. For some of us, the formulation of a valid and logically developed question is the major purpose of the assignment, and a highly detailed description of the methodology will not be required. For such an instructor, you may, for example, propose to extract and separate proteins without actually having to know in detail how this is accomplished. But other instructors may feel that your mastery or knowledge of methodological detail is as important as the validity of the questions posed. Both approaches are defensible, depending largely on the nature of the field of inquiry, on the level of the course being taken, and on the amount of laboratory experience you have had. Be sure you understand what your instructor expects of you before preparing this section of your paper.

As you describe each component of your proposed research, indicate clearly what specific question each experiment is designed to address, as in the following three examples:

To see if there is a seasonal difference in the amount of hormone present in the bag cells that induce egg-laying in Aplysia californica, bag cells will be dissected out of mature individuals each month and . . .

Before the influence of light intensity on the rate of photosynthesis can be documented, populations of the test species (wild columbine, Aquilegia canadensis) must be established in the laboratory. This will be done by . . .

To monitor seasonal changes in the relative abundance of macroalgae at different levels in the rocky intertidal zone at Nahant, MA, I will inspect each of the 10 boulders at monthly intervals for a 12-month period. At each inspection, I will . . .

Citing References and Preparing the Literature-Cited Section

Cite references directly in the text by author and year, as in the examples given earlier in this chapter (see also pp. 95–97). The Literature Cited section of your proposal is prepared as already described in Chapter 3 (pp. 120–123).

THE LIFE OF A REAL RESEARCH PROPOSAL

This is no idle exercise; the formal proposals written by practicing biologists are prepared exactly as described, except that each proposal must adhere strictly to the particular format (major headings, page length, type size, width of margins, number of copies to be submitted) requested by the National Science Foundation, National Institutes of Health, or other targeted funding agency. Proposals must be submitted by specified due dates or they will not be considered—no excuses are accepted. Copies of the proposal are then sent out to perhaps six to ten other biologists for anonymous reviews. A panel of still other biologists then meets to discuss the proposal and the reviews, and to make its own evaluation. If your case is well argued, it may receive funding; if it is not well argued, there is little hope. Learning to write a tightly organized and convincing proposal now will surely make your life easier later.

7
Writing for a General Audience: Science Journalism

Explaining scientific advances to the general public is a worthwhile—even a noble—endeavor, and if you have a substantial background in the subjects you are writing about you will have a real advantage over most journalists, who typically are trained in journalism rather than science. But the purpose of this chapter is not to prepare you for a career in newspaper or magazine publishing. Rather, I include this chapter because writing for a general audience can sharpen your thinking skills and reveal very clearly to your instructor how much you have really learned from what you have read.

SCIENCE JOURNALISM BASED ON PUBLISHED RESEARCH

Assimilating a piece of research published in the primary scientific literature and reorganizing that information to produce a successful piece of science journalism is an excellent exercise in summarizing information, simplifying complex material, and dejargonizing your writing. Most important, writing for a general audience of intelligent nonscientists is a wonderful way to tell if you really understand something. The science journalist is essentially a teacher. But as you may have already discovered, trying to teach something to someone else is one of the best ways of teaching it to yourself. It can also be fun—fun for you to write and fun for your instructor to read.

Science journalism differs from most of the other forms of writing discussed in this book in that the take-home message is always presented at the beginning of the article rather than at its end. That is, the article *begins* by summarizing what is to follow. By the end of the first or second short paragraph, for example, one typically finds sentences like these.

> Scientists have now found that lobsters use an internal magnetic compass to navigate during their annual mass migrations into deeper waters.
>
> In a recent report published in the journal Nature, Professors Tia-Lynn Ashman and Daniel J. Schoen present the remarkable finding that plants time their production of flowers in much the same way that people run efficient businesses.
>
> According to Professors Graziano Fiorito and Pietro Scutto, working at their laboratories in Italy, the common octopus can not only be trained to distinguish between objects of different colors, but can in fact learn to make these distinctions more quickly from each other than from human trainers.
>
> Researchers at the Dr. Seus School of Medicine and the Mt. Auburn Hospital have discovered an inherited molecular defect that makes some people naturally resistant to malaria, a disease affecting over 300 million people in tropical areas around the world.

Often such sentences begin the article. In other cases, summary sentences are preceded by one or a few sentences designed to stimulate additional reader interest in the topic. The opening sentences are known as the lead. Leads tend to follow one of four major formats: the simple statement, the bullet lead, the narrative lead, and the surprise or paradox lead. Some of these leads allow the writer considerable room for creativity.

The first is a simple, but dramatic statement of the major finding, usually in a single sentence, as in the example about malaria given before. A more interesting lead, but one that is more challenging to write, is called the bullet lead. Actually, it consists of three bullets, which are always followed by the general summary statement. For example:

> We all know people that have trained their dogs to fetch the daily newspaper without tearing it. Similarly, we all know that horses can be trained to respond to the slightest movement of their riders. And we all know that goldfish can be trained to come to the front of the fish bowl at the sound of a bell. Now it turns out that even the common octopus can be trained to perform certain simple tasks, and that they actually learn those tasks more quickly from each other than from a human trainer.

If the bullets are fired successfully, by the end of the third "shot," the reader is wondering how the individual bullets are related and where they are "leading." And just at that moment, the skillful writer answers those questions; if done properly, the reader wants to read more.

Another common lead takes the form of a narrative; it tells a story of some sort, and then follows up with the summary sentence.

> Sitting at the bottom of a large glass tank is a two-pound octopus. The octopus has been trained for several weeks to avoid balls of one color and to pick up balls of a different color. Every day for six hours he has been rewarded with food for choosing the right balls, and punished with mild electric shocks for choosing the wrong ones. Now, he sits idly in the tank, his eyes apparently following every movement of the researchers as they prepare to set up the next experiment, his mantle cavity filling and emptying in a consistent respiratory rhythm.
>
> The researchers bring over a tank containing another octopus, one that was freshly collected that morning from the warm and inviting waters just

outside the marine laboratory. The two octopi quickly crawl toward each other in their respective tanks, peering through the glass with apparent interest. "Now watch this," one of the researchers says to the newcomer, as she puts the trained octopus through his morning paces. The newly collected octopus watches, and seems genuinely interested in what the other octopus is doing. Now the researchers offer the same choices to the new octopus. Remarkably, after watching only four trials, the observing octopus chooses the correct ball over the other one in every one of the trials.

The surprising finding that octopi can learn from watching each other was recently published in the research journal Science by two biologists working at laboratories on the Italian coast, Professors Graziano Fiorito and Pietro Scotto.

Finally, there is the lead that tries to arouse the reader's attention by making a surprising or paradoxical statement, and then follows up with the summary sentence. Here is an example.

Biologists have for years spent many tedious hours training animals to perform simple tasks, by rewarding the desired behavior and punishing the undesired behavior. Now it seems that at least some animals may learn far more quickly by simply watching each other than by being trained by humans.

Two Italian scientists, Professor Graziano Fiorito and Professor Pietro Scotto announced in a recent issue of the research journal Science, that the common octopus can not only be trained to distinguish between objects of different colors, but can in fact learn to make these distinctions even more quickly by simply watching each other.

No matter how it begins, the rest of the article expands on the summary that has preceded it. For the article to be effective for a general audience, you must be careful to explain to the reader what was done, why it was done, what happened, and why the result is interesting, avoiding big words whenever possible and carefully explaining any terms that are essential to the story. Remember, you are not trying to impress or bamboozle others; you are trying to teach them something. Here is an example that further develops the story on octopus learning. If this was your own work, of course, you would submit it to your instructor double-spaced. Note that paragraphs tend to be shorter than in other forms of writing in Biology.

> Two Italian scientists, Professor Graziano Fiorito and Professor Pietro Scotto, announced in a recent issue of the research journal Science, that the common octopus can not only be trained to distinguish between objects of different colors, but can in fact learn to make these distinctions even more quickly by simply watching each other.
>
> They performed their study with Octopus vulgaris collected from the Bay of Naples, Italy. First they trained 30 individuals to grab a red plastic ball and 14 individuals to prefer a white plastic ball, by rewarding the animals with food if they chose correctly and mildly shocking the animals if they chose incorrectly. After 17-21 trials, the octopuses all learned to make the correct choice and no further training was necessary.
>
> There is nothing unexpected so far: biologists have been successfully training octopuses to exhibit simple behaviors for many decades. The surprising part of the experiment came when the researchers then let each trained octopus exhibit its acquired color preference four times to a freshly caught octopus that had received no prior training.
>
> The observing octopuses were then themselves given the opportunity to select either a white or a red plastic ball. Remarkably, all but a few of the 44 observers chose the color preferred by the octopus it had watched. Just as remarkably, the color preferences shown by the observing octopuses persisted for at least five days, when they were tested again.
>
> Clearly, the octopus is a quick learner. And it learns more quickly from simply watching what other octopuses do than it does by being shocked or rewarded by humans in the standard laboratory experiment. One wonders about the extent to which octo-

```
puses can learn more complex behaviors from each
other, and about the things octopuses might actu-
ally be learning by example in their natural envi-
ronment. Learning by example is apparently not a
uniquely vertebrate characteristic; octopus see,
octopus do.
```

JOURNALISM BASED ON AN INTERVIEW

An alternative to writing a piece of science journalism based on a published research article is writing a piece based on research that someone in your Biology Department is currently performing. Obviously, you need to get the professor's permission and cooperation to do this, and you have to do some preparation beforehand by reading some general textbook references about the research area being investigated and perhaps by looking over a few papers that the professor has recently published. Allow about one-half hour for the interview, and have at least a half-dozen questions prepared in advance. Here are some possible questions to get an interview started:

What basic question are you asking in your research?

How did you start doing research in this area?

What do you enjoy most about doing research?

What is the most surprising thing you have found out so far?

How did you find that out?

Although biologists are generally eager to explain what they do and why they do it, most of us don't get much practice talking about our research to undergraduates, so don't be surprised if the Biology professor you interview is difficult to follow at first. If you don't understand something that the professor says to you, don't be embarrassed or afraid to ask for clarification. Any lack of understanding is sure to be apparent in what you write; you can't explain something to someone else that you yourself don't understand. And in science journalism, unlike some other forms of writing in Biology courses, you can't hide your ignorance behind big words and convoluted sentence structures. The success of your writing depends heavily on how hard you work to understand your source. Along the way, you will learn a lot, meet some interesting people, and might even end up with something that you can publish in your campus newspaper.

One good way to prepare for the interview is to read examples of science journalism in major newspapers and magazines, such as the *New York Times* and *Newsweek.* For each article, ask yourself, "What questions did the interviewer probably ask his or her subject in conducting the interview?" and "What additional questions would I ask if I got to talk to this person myself someday?" That sort of thinking puts you in exactly the right frame of mind for the real thing.

Here is an example of interview-based science journalism. See if you can determine what questions the student asked the person he was interviewing.

> "Don't worry, I'm on the pill," he said, to allay her contraceptive concerns. Such male contraceptive pills may be available in the near future, resulting from research being conducted by Professor Norman Hecht at Tufts University.
>
> Over the past 20 years or so Professor Hecht has worked to understand how sperm develop, reasoning that the ability to block their normal development in the testis could lead to an effective contraceptive pill.
>
> As sperm develop within the testis, many new proteins must be synthesized in a particular order, or the sperm will not be able to function properly.
>
> Hecht and his coworkers—4 postdoctoral fellows, one graduate student, and collaborators in 5 other research laboratories around the world—have so far isolated 7 of these sperm-specific proteins and are now studying the expression of the genes responsi-

ble for producing them. What turns the expression of these genes on at the appropriate time in sperm development? What turns the expression of these genes off at the appropriate time in sperm development? If he can disrupt the normal expression of the genes coding for those proteins, abnormal—and ineffective—sperm should result.

Sperm production can be attacked at a number of points in their development (called spermatogenesis), by interfering with either the transcription of genes from DNA, the storage of the resulting gene messages (called messenger RNA, or mRNA for short), or the translation of those messages into the final proteins.

The genes coding for the testis-specific proteins that Professor Hecht is interested in are transcribed from their DNA templates early in spermatogenesis and are then stored in the cytoplasm for many days before being translated into proteins. "If we can prevent either the transcription of the genes or the subsequent translation of the mRNA's encoding these genes," says Professor Hecht with great enthusiasm, "we should be able to prevent the development of normal, functional sperm."

The great appeal of this approach, which Professor Hecht believes to be unique in the field, is that it should be possible to block sperm development without interfering with physiological processes elsewhere in the body.

One protein of particular interest to Professor Hecht is called contrin, which he and his coworkers discovered and isolated last year. Contrin binds to both DNA and mRNA. When it binds to DNA, it promotes transcription. Contrin also binds to mRNA, preparing it for long-term storage. Preventing germ cells from synthesizing contrin might therefore disrupt normal sperm development both by preventing certain transcriptional processes from occurring and by preventing the storage of mRNA molecules that are transcribed. No one yet knows how crucial contrin is for normal sperm development, but Professor Hecht expects to be the first to find out.

Professor Hecht has recently isolated the genes coding for mammalian contrin, and has successfully synthesized contrin in the laboratory using the cloned DNA sequences. With unlimited quantities of pure contrin in hand, the next step will be to determine the precise three-dimensional structure of the contrin molecule. It should then be possible to create artificially other molecules that bind specifically to contrin and disable it. Then it should be possible to prevent contrin synthesis in male mice and determine what happens to spermatogenesis in the absence of contrin.

This could be a first exciting step toward deliberately blocking the synthesis of crucial sperm-specific proteins indefinitely.

How much longer do men (and women) have to wait before male contraceptive pills become available commercially? "It will probably happen eventually," says Professor Hecht, "but not immediately."

"There are those in the United States and abroad," he says, "who believe the time and technology are right for widespread use of male contraceptives." It looks as though his laboratory will remain active for quite some time.

It is not difficult to see that the writer of this piece understands his subject well. And one certainly gets the impression that he enjoyed the assignment. In fact, I happen to know that he did.

8
Preparing Oral Presentations

Oral in-class presentations of published research papers are often assigned in conjunction with or in place of the written summaries or critiques discussed in the previous chapter. Research projects may also culminate in oral presentations. Although this book is about writing, I include this short chapter on talking because oral presentations are developed in much the same way as their written counterparts and can, in fact, provide an ideal framework for later expansion into written papers of any size. Most of the advice that follows applies to any sort of oral presentation; in writing any paper—summary, critique, research report, research proposal, or literature review—it typically helps to think first in terms of giving a clear talk.

Despite some major similarities, an oral presentation must differ from a written presentation in one important respect: a typewritten page can be read slowly and pondered, and can be reread as often as necessary, until all points are understood; an oral report, however, gives the listener only one chance to grasp the material. An analogy can be made with music. Before about 1910, music, to be successful, had to be liked at the first hearing; appreciation of subtleties might grow with repeated listenings, but composers knew that if their audience was not captivated by the first performance, that performance might well be the last. It was only with the invention of the phonograph that composers could sustain a career by intentionally delivering music intended to grow on its audience.

An oral presentation goes past the listener only once: for maximum impact, it must be very well organized, developed logically, stripped of details that divert the listener's attention from the essential points of the presentation, and delivered clearly and smoothly.

TALKING ABOUT PUBLISHED RESEARCH PAPERS

A talk, like any written work, can be effective only if you fully understand your topic. As suggested in earlier chapters, it is wise to skim the paper that you are presenting or discussing once or twice for general orientation, consult appropriate textbooks for background information as necessary, and pay particular attention to the Materials and Methods section and to the tables, graphs, and photographs included in the Results section. When you can summarize the essence of the paper in one or two sentences, you are ready to prepare your talk.

Preparing the Talk

The goal of your presentation is virtually identical with that of a written assignment: you seek to capture the essence of the research project—why it was undertaken, how it was undertaken, and what was learned—and to communicate that essence clearly, convincingly, and succinctly. Keep this goal in mind at all times.

1. **Do not simply paraphrase** the Introduction, Materials and Methods, Results, and Discussion sections of the paper or papers that you are presenting if you wish to keep your audience awake. To make an effective presentation, you must reorganize the information in each assigned paper. Begin your talk by providing background infor-

mation, drawing from the Introduction and Discussion sections of the paper and from outside sources if necessary so that the listener can appreciate why the study was undertaken. End your introduction with a concise statement of the specific question or questions addressed in the paper under discussion.

2. **Be selective; delete extraneous details.** Much of what is appropriate in a research paper is not necessarily appropriate for a talk about that paper. Since the listener has only one chance to get the point, some of the details in the paper—particularly methodological details—must be pruned out in preparing the oral presentation. Streamline; include only those details needed to understand what comes later. If, for example, you will not discuss the influence of sample size on the results obtained, do not burden the listener with such details in your talk. Similarly, is it important that the samples were mixed on a shaker table? If you never discuss this detail later in your talk, omit it at the outset. Tidbits such as these sometimes come out in the question period following your presentation but do not belong in the presentation itself.

3. **Focus your talk on the results.**

4. **Draw conclusions as you present each component of the study** so that you lead in logical fashion from one part of the study to the next. If you are discussing several experiments from a single paper, state the first specific question, briefly describe how it was addressed, present the key results, lead into the second specific question, describe how that question was addressed, present the key results, lead into the next question, and so forth. For ex-

ample: "The oyster larvae grew 20 μm/day when fed diet *A,* 25 μm/day when fed diet *B,* and 65 μm/day when fed a combination of diets *A* and *B.* This suggests that important nutrients missing in each individual diet were provided when the diets were used in combination. To determine what these missing nutrients might be. . . ." Lead your audience by the nose from point to point.

5. **Plan to use the blackboard or overhead transparencies.** A simple summary table or two is helpful when numbers are being discussed; numbers floating around in the air are difficult for listeners to keep track of. A diagram of experimental protocol can help the listener follow the plan of a study, along the lines of Figure 3.1 (p. 55), for example. Data can often be effectively summarized in a few graphs, even when those data were presented in the original paper as complicated tables. Keep the graphs simple, and be sure to label both axes. You need not reproduce graphs exactly as given in the paper, and you need not display every entry from a particular table. Focus on showing the trends in the data, and omit anything that fails to help you make your point clearly; I discuss this more fully at the end of the chapter.

6. **Summarize the major findings of the research at the end of your talk,** driving the points home one by one. You may wish to end your talk with a brief discussion of the way the study could be improved or expanded in the future, but don't set out to discredit the authors. End on a positive note, reinforcing what you want your audience to remember.

7. **Be prepared for questions about methodology.** Listeners often ask about interpretations of the data; to an-

swer these questions, you must be thoroughly familiar with the way the study was conducted.

Giving the Talk

1. **Know what you're going to say and how you're going to say it.** Hesitation, vagueness, and searching for words will all suggest a lack of understanding and will lose the attention of your audience. Write out your talk and practice it until you can produce a smooth delivery.

2. **Don't rush.** Write on the blackboard or on overhead transparencies during your presentation when, for example, labeling the axes of graphs. This helps punctuate your statements and also gives the listener time to digest what you are showing as well as time to take notes. For the same reason, you should label curves as you draw and talk about them. It is often a mistake to put completed illustrations on the blackboard ahead of time; the listener generally gets deprived of the opportunity to absorb what is being presented.

3. **Make the blackboard work for you** by drawing the listeners' attention to specific aspects of the graphs and tables that represent the point you wish to make. Don't simply say, "This is shown in the graph on the board." Rather, say "For example, all the animals fed on diets *A* and *B* grew at comparable rates, but . . ." and be sure to point to the data as you speak.

4. **Write unfamiliar terms on the blackboard or overhead transparency and avoid acronyms whenever possible;** there is no justification for referring to "NCAMs" instead of "neural cell adhesion molecules" when the term is used only once or twice in the talk. Remember, your goal is to communicate, not to impress or confuse.

5. **If using an overhead projector, point to the screen when you wish to highlight a detail, not to the transparency itself.** Transparency projectors magnify, and will transform the perfectly natural, barely noticeable nervous tremor of your hands into a highly distracting display of stage fright. It is not possible to seem at ease and self-confident when you appear to be experiencing an internal earthquake. Pointing to the screen, you are more likely to appear calm and collected.

6. **Don't mumble. Make eye contact with your listeners.** Don't talk to the blackboard.

7. **Try to sound interested in what you are saying,** no matter how many times you have practiced your talk. If you seem bored by your own presentation, you will most certainly bore the audience.

8. **Don't automatically refer to the author of a paper as *he*.** Many papers are written by women, and many are written by two or more researchers.

9. **Don't end abruptly.** Warn your audience when you are nearing the end of your talk by saying something like, "I would like to make one final point," or "Before I end, I wish to emphasize that. . . ." Such phrasings will prepare the listeners to receive your summary statements.

10. **End your talk gracefully.** A self-conscious giggle or a "Well, I guess that's it" isn't the best way to close an otherwise captivating presentation. I suggest something

like, "Thank you. I would be happy to answer any questions."

11. **Do not allow your presentation to exceed the time allotted.** You will lose considerable good will by rambling on beyond your time limit. Here again, a few practice sessions come in handy.

12. **Do not feel compelled to answer questions that you don't understand** during the question period after your presentation has been given. Politely ask for clarification until you figure out what is being asked.

13. **Paraphrase each question before answering it** so as not to lose the rest of the audience (and to buy yourself a few precious seconds to think). "The question is, does the technique used to isolate the DNA interfere with. . . ." Then address your answer to the entire audience, not just the person asking the question.

14. **Do not be afraid to admit that you don't know the answer to a question.** You can easily work your neck into a noose by pretending you know more than you really do; nobody expects you to be the world's authority on the topic you are presenting. Simply saying, "I don't know" is the safest way to go.

TALKING ABOUT ORIGINAL RESEARCH

Follow the format described for preparing and presenting your work. Again, begin by presenting the background information that listeners need to understand why you addressed the particular question or issue you chose to address, and then clearly state the specific question or issue being considered. Focus on the results of previous studies when presenting the background information and on your own results when giving the rest of the talk. Draw your conclusions point by point as you discuss each facet of the study, showing how each observation or experiment led to the next aspect of the work. End the talk by summarizing your major findings with their potential significance, and perhaps with a brief suggestion of what you might do next to further explore the issue you raised at the start of your talk.

TALKING ABOUT PROPOSED RESEARCH

This is similar to presenting a research paper except that you have more literature to review. Highlight a few key papers that show particularly clearly why the question you wish to address is a worthwhile and logical one, and again focus on the results of the studies you discuss. Then state the specific question you plan to address in your own work, being sure this question follows logically from the work you have just summarized. Finally, describe the approach you will take, focus on what you will do, and make clear what each step of the study is designed to accomplish. Conclude by briefly summarizing how the proposed work will address the question under consideration.

THE LISTENER'S RESPONSIBILITY

Few things in life are more disappointing than putting your heart and soul into preparing and delivering a talk to an audience that shows complete, apparent indifference. When you are a member of that audience, you bear a responsibility to listen closely, and to show the speaker that you listened closely. Try to formulate at least one question by the end of the talk, about something you didn't understand, something you thought was particularly interesting ("I was amazed by the ability of those polar fishes to keep from freezing. Do local fish produce the same kind of biological antifreezes in the winter?"), or something unusual you saw in the data ("In that table you showed us, why were the arsenic concentrations so high in the control animals' tissues?").

Even if the speaker can't answer your question, he or she will at least detect some interest in the talk and feel flattered that you cared enough about his or her development as a biologist and seminar speaker to have paid so much attention.

PREPARING EFFECTIVE SLIDES AND OVERHEADS

A figure or table that works fine in a published paper may not work nearly as well as a visual aid for a talk. When reading a journal article, readers can scrutinize your data as long as required, over several cups of coffee if necessary. For a talk, however, you want the audience to understand the slide or overhead quickly so that you can concentrate on the results. If the slide is too complicated or too difficult to read, you may be finished talking about the results while your listeners are still trying to figure out what your axes are! Make your visual aids as simple and as clear as possible. They should ease communication, not hinder it.

Consider Figure 8.1, which was developed for publication. It concerns seasonal variation in initial carbon content and juvenile growth rate of an intertidal barnacle, *Semibalanus balanoides.* Combined with its caption, the figure is a perfectly acceptable, self-sufficient summary of the data. But it would not work well as

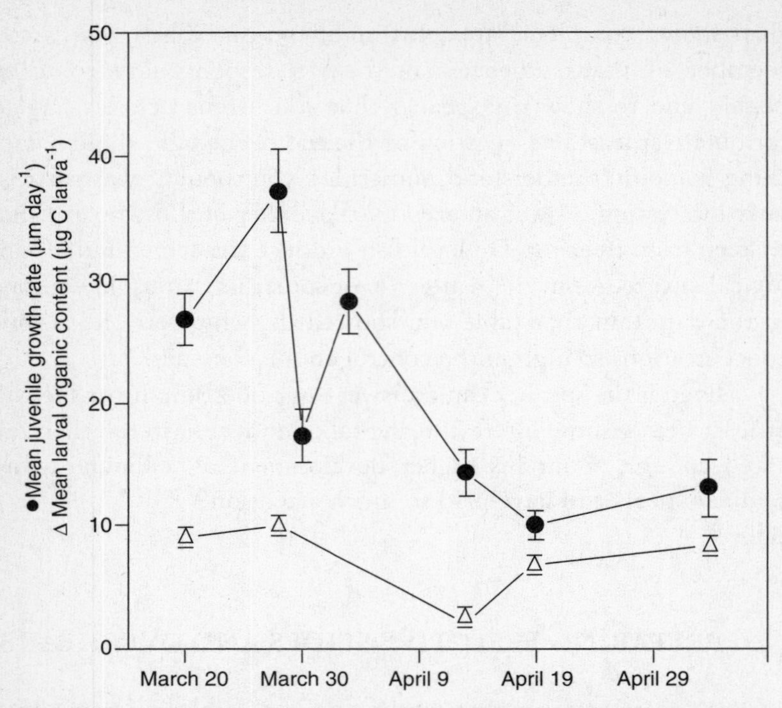

Figure 1. Temporal variation in larval organic content and juvenile growth rate of the barnacle, Semibalanus balanoides. Larvae that attached to artificial substrates in the low intertidal zone were collected in the field at intervals during 1995; individual organic content was estimated by dichromate oxidation. Metamorphosed juveniles collected from the field were reared for 5–7 days in the laboratory under controlled conditions to determine growth rate. Each point is the mean (± one standard error) of 13–33 individual measurements.

Figure 8.1
A graph with its figure caption, designed for publication in a formal research paper. (Courtesy of J. Jarrett)

a slide. For one thing, the lettering is too small; people sitting more than a few rows from the front of the room would likely be unable to read the axis labels, and they certainly wouldn't be able to read the caption. And you don't want them reading the caption while you're talking anyway: you want them to be listening to what you are saying. What's more, the listener must look back and forth to the left of the slide to interpret what the two curves represent.

Figure 8.2 shows how this presentation might be modified for conversion to an effective slide or overhead. By using two lines instead of one for the dates (*X*-axis) and the *Y*-axis titles, and by shifting the *Y*-axis titles to separate sides of the graph, we can use a much larger, more readable typeface. I have also identified the

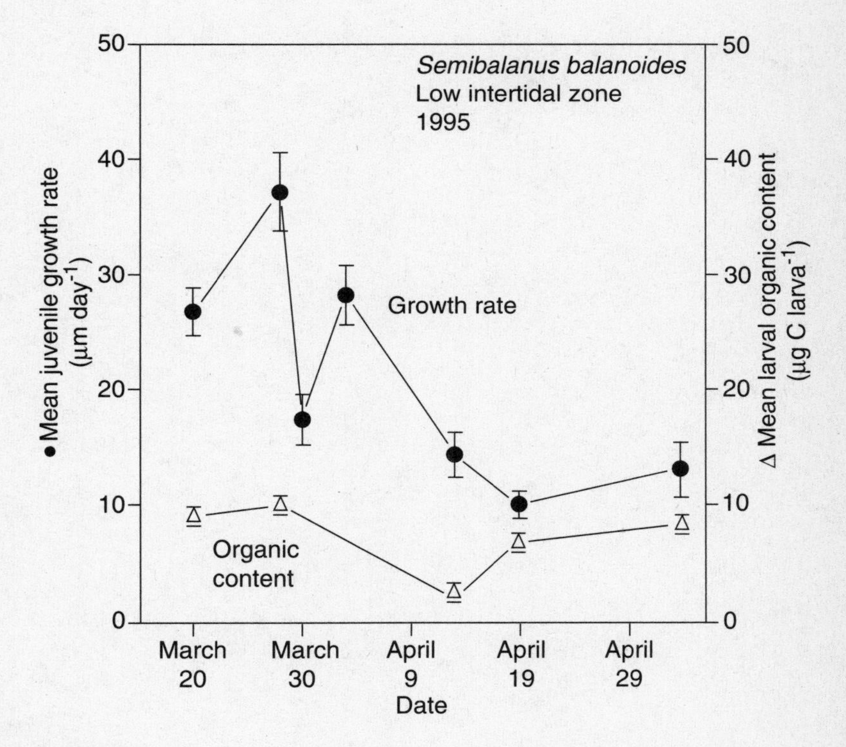

Figure 8.2
A revised version of Figure 8.1, designed for use as a slide or overhead during an oral presentation. Note that some of the information contained in the figure caption of Figure 8.1 is now placed directly on the graph. (Courtesy of J. Jarrett)

two curves directly on the graph, and have added the name of the species and where and when the data were collected. Finally, the figure caption has been removed. A speaker will have a far easier time orienting his or her listeners to a slide or overhead made from this figure. The audience should have little difficulty reading and understanding the illustrated data even if they are sitting in the back row. In designing slides and overheads, always try to reach the person sitting at the back of the room.

9
Writing Letters of Application

An application for a job, or for admission to a graduate or professional program, will generally include a résumé and accompanying cover letter (both of which you write), and several letters of recommendation (which you generally never get to see). When applying to graduate or professional schools, and often when applying for jobs, you will also include a transcript of your college coursework and any special examination scores—for example, Graduate Record Examination (GRE) scores. You have no control over what your transcript and GRE scores say about you; what is done is done. But you can still influence the message transmitted through your résumé and supporting letters.

Your résumé summarizes your educational background, relevant work experience, relevant research experience, goals, and general interests. The accompanying cover letter identifies the position for which you are applying and draws the reader's attention to those aspects of your résumé that make you a particularly worthy candidate. The recommendations will give an honest assessment of your strengths and weaknesses (we all have some of each) and offer the reader an image of you as a person and as a potential employee or participant in a professional program. In this chapter, I will consider the art of preparing effective résumés and cover letters, and of increasing the odds of ending up with effective letters of recommendation.

BEFORE YOU START

Always try to put yourself in the position of the people who will be reading your application. What will they be looking for? They will probably be considering your application with three main questions in mind:

1. Is the applicant qualified for this particular position?
2. Is the applicant really interested in our program or company?
3. Will the applicant fit in here?

Your application must address all three issues.

When you prepare your application, you should also consider that the number of applications received by a potential employer or professional school usually exceeds the number of positions available, and often by a considerable margin. Many applicants will be qualified for the position, yet not every applicant can be interviewed or offered admission. Whoever begins reading your application will necessarily be looking for any excuse to disqualify you from the competition; your goal, then, must be to hold the reader's interest to the end.

PREPARING THE RÉSUMÉ

The people who read your application will not spend hours scrutinizing your résumé; probably they will examine it for only one or two minutes at most. An effective résumé is therefore well organized, neat, and as brief as possible. Your résumé should not exceed two pages in length.

There is no standard format for a résumé; the model given in Figure 9.1 should be modified in any way that emphasizes your particular strengths and satisfies your own esthetic sense. The résumé is, however, no place to be artsy or cute; don't do anything that might suggest that you are not taking the application process seriously.

Ruth Ray Parker

Address and phone number
Until June 1, 1996:
P.O. Box 029
University Station
Kingston, RI 02881
(401) 201-1717

After June 1:
29 Lakeview Drive
Narragansett, RI 02882
(401) 788-0153

Date of Birth: September 13, 1978

Goals: To earn a Ph.D. in Comparative Immunology and pursue a career in teaching and research.

Education

University of Rhode Island, Fall 1992—Spring 1996.
Major: Biology

Research Experience

Conducted a one-semester research project (in Dr. Oliver Hornbeam's laboratory) on the structure and function of guard cells in lyre-leaved sage, *Salvia lyrata* L., using transmission electron microscopy. Presented the results of ths research at the 38th New England Undergraduate Research Conference, Siegel University, Barnum, CT (May 1995).

Teaching Experience

Undergraduate teaching assistant for introductory Biology laboratory, Fall semester 1995.

Honors

Elected to Phi Beta Kappa honor society, Spring 1995.
Received Churchill Prize in Biology (for performance in introductory Biology course), Spring 1993.
Dean's List seven out of eight semesters.
Selected for teaching assistant position noted above.

```
Work Experience

  Summer 1993. Counselor, Lake Baker Summer Camp, AK
  Summer 1994, 1995. Worked for Sweet Pea's Garden
    Center, Falmouth MA. Cared for all plants and shrubs,
    with one assistant.
  Summer 1996. Assisted in the culture of oysters and
    hard shell clams, Mook Sea Farms, Damariscotta, ME.

Special Skills

  Tissue preparation (fixation, embedding, sectioning)
    for transmission electron microscopy.
  Operation of JEOL Model 100CX transmission electron
    microscope
  Developing 35-mm black and white film and TEM
    negatives.
  Printing black and white photographs.

Outside Activities

  Saxaphonist, URI Jazz band, 1994-1996
  Swim team (1994-1996, captain 1996)
  Campus tour guide Fall 1994, Fall 1995
  Vegetable gardening (each summer since 1991)
```

Figure 9.1
Sample résumé.

All résumés must contain the following three components:

1. Full name, address, telephone number (and e-mail address if you have one)
2. Educational history
3. Relevant work, teaching, and research experience, if any

In addition, you will want to add any other information that makes you look talented or well-rounded or both.

4. Honors received
5. Papers published
6. Special skills
7. Outside activities, sports, hobbies

Avoid drawing attention here to any potential weaknesses; if, for example, you lack teaching experience, do not write "Teaching experience: none." Use the résumé exclusively to play up your strengths.

You might also add a one- or two-sentence statement of your immediate and long-range goals, if known, and the names of people who have agreed to write references on your behalf; this material is often incorporated into the cover letter instead, as discussed shortly.

You are not required to list age, race, marital status, height, weight, sex, or any other personal characteristic. Be self-serving in deciding what to include. If you think your youth might put you at a disadvantage, omit this information. If you think your age, race, or sex might give you a slight competitive edge, by all means include that information.

Do not be concerned if your first résumé looks skimpy; it will fill out as the years go by. It is better to present a short, concise résumé than an obviously padded one.

You should alter your résumé for each application completed to focus on the different strengths required by different jobs or programs. If, for example, Ruth Ray Parker, whose résumé appears in Figure 9.1, were applying to a marine underwater research program, she might add under Special Skills, that she is a certified SCUBA diver. If she were applying for a laboratory job in a hospital, she would probably omit the information about diving certification.

PREPARING THE COVER LETTER

The cover letter plays a large role in the application process and is usually the first part of your application read by an admissions committee or prospective employer. A well-crafted letter of application can do much to counteract a mediocre academic record. A poorly crafted letter, on the other hand, can do much to annihilate the good impression made by a strong academic performance. Keep revising this letter until you know it works well on your behalf. Have some friends, or perhaps an instructor, read and comment on your letter; then revise it again. Be sure to type or computer-print the final copy; neatness counts, and typing also conveys seriousness of purpose. The time you put into polishing your cover letter is time well spent. The cover letter should be about one or, at most, two typed pages.

Do not simply write,

> Dear M. Pasteur:
> I am applying for the position advertised in the Boston Globe. My resumé is enclosed. Thank you for your consideration.
>
> Sincerely,
>
> *Earl N. Meyer*
>
> Earl N. Meyer

Although the letter ends well, its beginning is vague and its midsection does little to further the applicant's cause. Use the cover letter to

1. Identify the specific position for which you are applying (Monsieur Pasteur may have several positions open. Earl N. Meyer is applying for the position of research assistant, but how is M. Pasteur to know?)
2. Draw the reader's attention to those elements of your résumé that you feel make you a particularly qualified candidate
3. Indicate that you understand what the position entails and that you have the skills necessary to do a good job
4. Convince the reader you are a mature, responsible person
5. Convey a genuine sense of enthusiasm and motivation

Before you begin to write the letter, ask yourself some difficult questions, and jot down some carefully considered answers:

> Why do I want this particular job or to enter this particular graduate program?
>
> What skills would be most useful in such a job or program?

Which of these skills do I have?

What evidence of these skills can I present?

Your answers to these questions will provide the pattern and the yarn from which you will weave your cover letter.

Not everyone will have a résumé that looks like Ruth Parker's (Figure 9.1). But you need not have made the Dean's List every semester or have had formal teaching or research experience in order to impress someone with your application. In your cover letter, focus on the experiences that you *have* had. In lieu of teaching, perhaps you have done formal or informal tutoring. In lieu of having had formal research experience, perhaps you have taken numerous laboratory courses. Perhaps some experience you had in one or more of these laboratory courses influenced your decision to apply to a particular job or program. Perhaps acquiring certain skills in one or more of these laboratory courses has prepared you for the program or position to which you are applying. Or perhaps you can draw from experience outside Biology to document reliability, desire, and willingness to learn new things, or ability to learn new techniques quickly. We all have strengths; decide what yours are and which ones are appropriate for inclusion in your application.

Tailor each letter to the particular position or program for which it is being prepared. Try to find some special reason for applying to each program; if possible, your application should reflect deliberate choice and a clear sense of purpose. If, for example, you have read papers written by a faculty member at the institution to which you are applying and have become interested in that person's research, weave this information into your cover letter. Should you take this approach, you must say enough about that person's research or research area to make clear that you understand what you are writing about. On the other hand, if your major reason for wanting a particular job or wanting to attend a particular graduate program is the geographical location of the company or school, be careful not to state this as your sole reason for applying; as you write, and as you reread what you have written, try to put yourself on the receiving end of the cover letter and consider how your statements might be interpreted.

Back up all statements with supporting details. Avoid simply saying that you have considerable research experience. Instead, briefly explain what your research experience has been. Do not state that you are a gifted teacher; describe your teaching experience. State the facts and let the reader draw the proper inferences.

Sign the letter with your given name, not a nickname; again, don't run the risk of not being taken seriously.

Here is an example of a weak cover letter. Similar letters have, unfortunately, been submitted by people with very good grades, test scores, and letters of recommendation. The author is applying for admission to a Ph.D. program in Biology.

To the admissions committee:

I have always been fascinated by the living world around me. I marvel at the details of the way Biology works, and I would now like to fulfill my curiosity and passion for Biology in pursuing a Ph.D. in your program.

As you can see from my transcript, I have taken twelve courses in Biology (two more than the number needed for graduation) and have done well in most of them. I am especially interested in plant physiology and did a one-semester research project on this subject during my senior year.

I would also like to apply for the teaching assistantship award. I have always liked helping people learn about science, and I am eager to communicate my enthusiasm for Biology to others. I have requested that my GRE scores be sent directly to you.

I look forward to your reply.

Sincerely,

Dynein Arms

Dynein Arms

Remember, the admissions committee is looking for any excuse to disqualify applicants; this letter may give the committee just that excuse regardless of what the rest of the application looks like. The letter does convey enthusiasm, but it is a very naive enthusiasm. What does the student find interesting about the physiology of plants? What was the research project? What question was asked? How was the question addressed? What results were obtained? Did the student learn anything from the experience? Does the student really know anything about plant physiology? What makes the student think that she would be an effective teacher? Has she had any teaching experience? Does she understand what teaching entails?

The rest of the application letter is equally uninformative. Why is the student applying to this particular program? All Biology majors take Biology courses, and the student's grades are already on the transcript. What has the student learned from these courses that makes her want to pursue advanced study?

The same student could have written a much more effective letter by thinking about what admissions committees might be looking for and by documenting her strengths. Here is an example of the way this student might have rewritten her letter.*

> To the admissions committee:
>
> Please consider my application for admission to your Ph.D. program in Biology. I will be graduating from Cerebral University in May with a B.S. in Botany. I believe that I have the experience and motivation to make a contribution to your program.
>
> I became interested in plant physiology through a seminar course taught by Professor Mendel. This was my first experience reading the original scientific literature, but by the end of the semester I was able to present a well-received research proposal on the subject of root growth, based largely upon my own library research.

*The research described in her letter is based on a paper by Kapulnik et al., 1985. *Canadian Journal of Botany* 63: 627–631.

Dr. Mendel later invited me to conduct, in his laboratory, a research project examining the influence of nitrogen-fixing bacteria (Azospirillum spp.) on wheat seedling (Triticum aestivum) root development. Briefly, we first surface-treated wheat seeds in three percent sodium hypochlorite, to kill adhering microorganisms, and then incubated the seeds on moist filter paper in sterile, covered petri dishes. After the seeds germinated, some of the plates were inoculated with nitrogen-fixing bacteria and others were left untreated as controls.

After the seedlings had incubated for seven days at 25°C, I measured maximum seedling length (at 12× using a dissecting microscope fitted with a calibrated ocular micrometer), average seedling live weight (using a Cahn electrobalance precise to 0.1 µg), and average effective surface area of the root system (based on rate of HCl uptake, using the titration method of Carley and Watson, 1966). The presence of nitrogen-fixing bacteria significantly enhanced root development ($P < 0.05$, t-test) and shoot development ($P < 0.05$, t-test). We still do not know whether increased plant growth resulted directly from the presence of the bacteria (possibly through bacterial secretion of growth-stimulating chemicals); the root may be stimulated by the bacteria to release a growth-promoting hormone of its own.

Through this study, I learned the importance of careful experimental design and, more important, that I have the patience to do research. I do not wish to commit myself to a specific field of research at this time, believing that I would benefit from an additional year of literature and laboratory exploration, but I believe that I would like to study the role of hormones in early plant development.

During my last semester at Cerebral University, I have been acting as an undergraduate teaching assistant for the introductory Biology laboratory. I find that having to explain things to other students forces me to come to grips with what I do and do not know. I am enjoying the challenge greatly,

```
and I look forward to doing additional teaching in
the future; I am learning a lot about Biology
through teaching.
    I have asked the following faculty members at
Cerebral University for letters of recommendation:
    Professor G. Mendel (Biology Dept.)
    Professor R. Chase (Biology Dept.)
    Professor W. Layton (Classics Dept.)
    Thank you for considering my application. I
look forward to receiving your response.
                                        Sincerely,
                                       Dynein Arms
                                       Dynein Arms
```

This letter, too, conveys enthusiasm, but it is an enthusiasm that reflects knowledge, experience, and commitment; the applicant seems to understand what research is all about and apparently knows how to go about doing it. Moreover, we see that Ms. Arms thinks clearly and writes well.

Dynein Arms would write a somewhat different letter if she were applying for a job as a technician in a research laboratory. In this letter, she would want to emphasize her skills and reliability as a laboratory worker and her interest in the type of research being done in the laboratory to which she is applying. An example of such a letter follows.

```
Dear Professor Hornbeam:
    Please consider my application for the technical
position you advertised recently in the Boston
Globe. I will be graduating from Cerebral University
this May with a B.S. degree in Botany. Although I
eventually expect to return to school to pursue a
Ph.D., I would first like to work in a plant physi-
ology laboratory for one or two years to learn some
additional techniques and to become more familiar
with various research fields and approaches.
    I first became interested in plant physiology
through a seminar course with Professor G. Mendel
at Cerebral University. During the semester, we
read two of your recent papers describing studies
```

on the hormonal control of Selaginella kraussiana orientation to light.

Following this seminar, I began doing research on wheat seedlings (Triticum aestivum) in Professor Mendel's laboratory, investigating the influence of nitrogen-fixing bacteria (Azospirillum spp.) on root development. We found that root development was significantly enhanced by the presence of nitrogen-fixing bacteria. The next step in the study will be to determine whether increased root growth results directly from the presence of the bacteria (possibly through bacterial secretion of growth-stimulating chemicals), or indirectly, by stimulation of the root to secrete a growth-promoting hormone.

Through this study, I learned a variety of general laboratory techniques (use of assorted balances, sterile culture methodology, measurement of effective root surface area using the titration technique of Carley and Watson, 1966), and I also learned that I have the patience and motivation needed to do careful research. I have taken five laboratory courses in Biology (General Genetics, Invertebrate Zoology, Comparative Animal Physiology, Cell Biology, and Developmental Biology), in which I learned several specialized laboratory techniques, including the pouring and use of electrophoretic gels, measurement of organismal and mitochondrial respiration rates, and use of a vapor pressure osmometer.

In short, I am very much interested in your research and believe I can make a contribution to the work of your laboratory. I have requested letters of recommendation from the following faculty at Cerebral University:

Professor G. Mendel (Biology Dept.)
Professor P.-Y. Qian (Biology Dept.)
Professor J. Phillips (Classics Dept.)

Thank you for considering my application; I look forward to hearing from you.

Sincerely,

Dynein Arms

Dynein Arms

Both of these letters convey knowledge of the position or program applied for, a sincere interest in Biology, and a high level of ability and commitment. Your letter should do the same. Your credentials may not be as impressive as Ms. Arms's, but if you think about the experience you have had in relationship to the skills required for the position or program, you should be able to construct an effective letter. Take your letter through several drafts until you get it right (see Chapter 10, on revising).

RECRUITING EFFECTIVE LETTERS OF RECOMMENDATION

Letters of recommendation can be extremely important in determining the fate of your application. Although you do not write these letters yourself and rarely even get the opportunity to read them, you can take steps to increase their effectiveness.

Getting an *A* in a course does not guarantee a strong letter of recommendation from the instructor of that course. The most useful letters to admissions committees and prospective employers are those commenting on such characteristics as the following: laboratory skills, communication skills (written and oral), motivation, ability to use time efficiently, curiosity, maturity, intelligence, ability to work independently, and ability to work with others. Instructors cannot comment on these attributes unless you become more than a grade in their record books. Make an appointment to talk with some of your instructors about your interests and plans. We faculty members are usually happy for the opportunity to get to know students better.

When it is time to request letters of recommendation, choose three or four instructors who know something about your abilities and goals, and ask each of them if he or she would be able to support your application by writing a letter of recommendation. Give each person the opportunity to decline your invitation. If the people you ask agree to write on your behalf, make their task easier by giving them a copy of your résumé, transcript, and letter of application and, if appropriate, a copy of the job advertisement. **Be certain to indicate clearly the application deadline** and the address to which the recommendation should be sent.

It takes as much time and thought to write an effective letter of recommendation as it takes to write an effective letter of application. Don't lose good will by requesting letters at the last minute. **Give your instructors at least two weeks** to work on these letters. "It has to be in by this Friday" will probably annoy your prospective advocate and may not allow the recommender the time needed to prepare a good letter even if he or she is still in a cooperative mood. Moreover, last-minute requests don't speak favorably about your planning and organizing abilities, and they imply a lack of respect for your instructor. So be considerate, and thereby get the best recommendation possible.

10
Revising

The preceding chapters concern the reading, note-taking, thinking, synthesizing, and organizing that permit you to capture your thoughts and your evidence in a first draft. This chapter concerns the revising that must follow, in which you examine the first draft critically and diagnose and treat the patient as necessary. I typically revise my own writing four or five times before letting anyone else see it and several more times after it has been reviewed by others, so don't feel inadequate for not producing flawless prose on your first or second draft. Good writers aren't necessarily more intelligent than you are; most of them just revise more often.

Writing a first draft gives you the opportunity to get facts, ideas, and phrasings on paper, where they won't escape. Once you have captured your thoughts, you can concentrate on reorganizing and rephrasing them in the clearest, most logical way. All writing benefits from revision. For one thing, the acts of writing and rereading what you have written typically clarify your thinking. Then, too, there is the universal difficulty in getting any point across (intact) to a reader, even when you finally know precisely what it is that you *want* to say. Revising your work improves communication and often leads you to a firmer understanding of what you are writing about.

It is difficult to revise your own work effectively unless you can examine it with a fresh eye. After all, you know what you wanted to say; without some distance from the work, you can't really tell whether or not you've actually said it. For this reason, plan to complete your first draft at least five days before the final product is due so as to allow time for careful revision. Reading your paper aloud—

and listening to yourself as you read—often reveals weaknesses that you would otherwise miss. It also helps to have one or more fellow students carefully read and comment on your draft at this stage of its development; it is always easier to identify writing problems—wordiness, ambiguity, faulty logic, faulty organization, spelling and grammatical errors—in the work of others, so forming a peer-editing group is a clear step toward more effective writing. Be sure to tell readers of your work that you sincerely want constructive criticism, not a pat on the back. Remember, your goal as a writer is to communicate—clearly and succinctly, making it as easy as possible for the reader to follow your argument. Your goal as a reader of someone else's draft is to help its author to do the same. At the end of this chapter, you will find advice on how to be an effective reviewer.

Choose whatever system works best for you, but always revise your papers before submitting them. No matter how sound, or even brilliant, your thoughts and arguments are, it is the manner in which you express them that will determine whether or not they are understood and appreciated (or, in later life, whether they are even read). With pencil or pen at the ready (and scissors and tape, too, if you are not using a computer), the time has come to edit your first draft: for content, for clarity, for conciseness, for flow (coherence), and for spelling and grammar. If you are writing with a word processor, make your revisions on printed copy rather than on-screen; to edit effectively, you must see more than one screen of text at a time. Continue editing and revising—printout by printout—until your work is ready for the eyes of the instructor, admissions committee, or potential employer. This chapter should help you know when you have arrived at that point.

REVISING FOR CONTENT

1. **Make sure every sentence says something.** Consider the following opening sentence for an essay on the tolerance of estuarine fish to changes in salinity:

```
Salinity is a very important factor in marine
environments.
```

What does this sentence say? Does the author really think readers need to be told that the ocean is salty? What *is* important about salinity? A careful editor will delete the sentence and begin anew with a sentence that says something worth reading. For example,

```
Estuarine fish may be subjected to enormous
changes in salinity within only a few hours.
```

Similarly, a sentence like

```
There are many physical and biological factors
that affect the growth of insect populations.
```

could profitably be revised to read

```
Growth rates of insect populations are influ-
enced by such environmental factors as temper-
ature, food supply, buildup of metabolic
wastes, availability of mates, and magnitude
of predation pressure.
```

The authors of the revised opening sentences know where their essays are headed, and so does the reader. The original versions got the writers started; the revision process focused the writers' attention on a destination.

2. **Use the word *relatively* only when making an explicit comparison.** Consider this example: "Many of the animals living near deep-sea hydrothermal vents are relatively large." The thoughtful reader wonders, "relative to what?" Either delete the word and replace it with

something of substance (for example, "Animals living near deep-sea hydrothermal vents can exceed lengths of three meters") or make a real comparison (for example, "Many of the animals living near deep-sea hydrothermal vents are large relative to their shallow-water counterparts," or "Some animals living near deep-sea hydrothermal vents are many times larger than their shallow-water counterparts").

3. **Never tell a reader that something is interesting.** Let the reader be the judge. Consider this rather uninformative sentence:

   `Cell death is a particularly interesting phenomenon.`

   Is the phenomenon interesting? If so, ask yourself *why* you find it interesting, and then make a statement that will interest the reader. This example could, for instance, be rewritten as follows:

   `During the development of all animals, certain cells are genetically programmed for an early death.`

4. **Be cautious in drawing conclusions, but not overly so.** It is always wise to be careful when interpreting biological data, particularly with access to only few experiments or small data sets. For instance, write "These data suggest that . . ." rather than "These data demonstrate, or prove that. . . ." But don't get carried away, as in the following example:

   `This suggests the possibility that inductive interactions between cells may be required for the differentiation of nerve tissue.`

Here, the author hedges three times in one sentence, using the words *suggests, possibility,* and *may.* Never hedge more than once per sentence, as in the following rewrite:

```
This suggests that inductive interactions are re-
quired for the differentiation of nerve tissue.
```

If you are too unsure of your opinion to write such a sentence, reexamine your opinion.

5. **While revising for content, keep in mind an audience of your peers, not your instructor.** In particular, be sure to define all scientific terms and abbreviations; it is not enough simply to use them properly. Brief definitions will help keep the attention of readers who may not know or may not remember the meaning of some terms and will also demonstrate to your instructor that you know the meaning of the specialized terminology you are using. Try to make your writing self-sufficient; the reader should not have to consult textbooks or other sources in order to understand what you are saying. As always, if you write so that you will understand your work years in the future, or so that your classmates will understand the work now, your papers and reports will generally have greater impact and will usually earn a higher grade.

REVISING FOR CLARITY

Be sure each sentence says what it's supposed to say; you want the reader's head to be nodding up and down, not side to side. Which way is the reader's head going in the following example?

```
These methods have different resorption rates and
tail shapes.
```

Do methods have tails? Can methods be resorbed? This sentence fails to communicate what its author had in mind. Indeed, it is difficult to tell *what* the author had in mind. Here is another sentence that does not reflect the intentions of its author.

> `From observations made in aquaria, feeding rates of the fish were highest at night.`

How many observers do you suppose can fit into an aquarium? Aquaria usually contain fish, not authors; is the author of our example all wet? A revised sentence might read,

> `Feeding rates of fish held in aquaria were highest at night.`

Some biologists are clearly more dedicated to their research than most of us are.

> `Ferguson (1963) examined autoradiographs of sea star digestive tissue after being fed radioactive clams.`

Perhaps we should feed the clams not to Ferguson but to the sea stars?

> `Ferguson (1963) fed radioactive clams to sea stars and then examined autoradiographs of the sea star digestive tissue.`

Note in the preceding example the advantages of summarizing a study in the order in which steps were undertaken; grammatical difficulties typically vanish, and the sentence automatically becomes clearer.

Here is another sentence that simply is not doing its author's bidding.

```
In order to keep the size of the samples constant,
the sampling pipet was calibrated so that the vol-
ume of a single drop was known.
```

It is difficult to see how calibrating a pipet will keep sample sizes constant. Presumably, the author means that the same pipet was used for each sample (which would keep sample sizes constant). That the size of each sample was known is really a separate, independent thought.

Confusing sentences inevitably arise when three or more nouns are lined up in a row. Consider this example.

```
Sleep study results show that tryptophan signifi-
cantly decreases the time needed to fall asleep
(Miller and Brown, 1991).
```

At the first reading, the reader probably expects "results" to be a verb, but instead it is a noun, preceded by two other nouns. The reader must stop and decode the sentence. Ah! The author is discussing the results of studies of people sleeping. We can rewrite the sentence to make this much clearer.

```
Recent studies show that tryptophan decreases the
time needed for people to fall asleep (Miller and
Brown, 1991).
```

In revising your work, think twice before leaving more than two nouns together; two is company, three is a crowd.

Here are two additional examples of unclear writing.

1. This determination was based on mannitol's relative toxicity to sodium chloride.
2. The surface area of mammalian small intestines is three to seven times greater than reptiles.

How can one chemical be toxic to another chemical? The author is probably trying to tell us that two chemicals differ in their toxicity to some organisms or cell types. With the second example, one wonders how an intestinal surface area can be greater than a reptile; again, the author is not making the comparison he or she intended. Readers should never have to guess what the proper comparison is; readers will appreciate your writing more the less you make them work. In any event, never invoke the "You know what I mean" defense. If a student writes, "A normal human fetus has 46 chromosomes," how can I assume the student understands that each *cell* of the fetus has 46 chromosomes? It is your job to inform the reader, never the reader's job to guess what you are trying to say.

Of course, it doesn't help that confusing sentences surround us in our everyday worlds. Consider this example taken from the local newspaper.

> `Offer void where prohibited by law, or while supplies last.`

The meaning of this sentence is not immediately clear. It is apparently impossible for anyone to take advantage of this offer; the offer is either prohibited by law or, if permitted by law, is void while supplies last. Supplies can never run out since the advertiser apparently is unwilling to fill your order as long as the items are available. If stocks become depleted, perhaps by eventual disintegration of the product, the advertiser could then honor your request; but the company would no longer have anything to send you!

Then there is this unfortunate wording I noticed on the side of some paper cups: "Put litter in its place." Since the cup isn't litter until it's on the ground, this is an apparent order to litter, probably not what the restaurant management intended. And how about this gem: "Our nuclear reactors are as safe as they can possibly be. And we are constantly working to make them safer."

With practice, you can find similarly confusing or absurd sentences almost anywhere you look; if they appear in your own writing, revise them. Make each sentence state its case unambiguously. Here is a sentence that does not do so.

> `Sea stars prey on a wide range of intertidal animals, depending on their size.`

Is the author talking about the size of the sea stars that are preying or about the size of the intertidal animals that are preyed upon? Don't be embarrassed at finding sentences like this one in early drafts of your papers and reports. Be embarrassed only when you don't edit them out of your final draft.

The Dangers of "It"

Frequent use of the pronouns *it, they, these, their, this,* and *them* in your writing should sound an alarm: probable ambiguity ahead. Consider the following example of the trouble *it* can cause:

> `The body is covered by a cuticle, but it is unwaxed.`

Which is unwaxed: the body or the cuticle? Similarly, *it* makes the second part of the following sentence equally ambiguous:

> `The chemical signal must then be transported to the specific target tissue, but it is effective only if it possesses appropriate receptors.`

Are these receptors needed by the chemical signal or by the target tissue? I'm confused. In the next example, *these* causes similar problems for the reader.

> `Antigens encounter lymphocytes in the spleen, tonsils, and other secondary lymphoid organs. These`

> then proliferate and differentiate into fully mature, antigen-specific effector cells.

Presumably the lymphocytes are proliferating, not the tonsils, although the author has certainly not made this clear. The problem is easily repaired by beginning the second sentence with "The lymphocytes. . . ." In the next example, *their* is guilty of a similar offense.

> Like fanworms and earthworms, leeches have proven very useful to neurophysiologists. Their neurons are few and large, making them particularly easy to study with electrodes.

Most readers are likely surprised to learn that neurophysiologists have so few neurons and are so easy to study. Now let us consider the difficulties *they* can cause.

> Tropical countries are home to both venomous and nonvenomous snakes. They kill their prey by constriction or by biting and swallowing them.

How much clearer the last sentence could become by replacing *they* with a few words of substance and by deleting *them* entirely:

> Tropical countries are home to both venomous and nonvenomous snakes, The nonvenomous snakes kill their prey by constriction or by biting and swallowing.

One more example will show just how troublesome *they* can be.

```
When the larval stage of the parasitic worm was ex-
posed only to animals of a species that never
serves as a host, they did not parasitize them.
```

If *they* have their way, the reader must guess who is not parasitizing whom. Realizing that the sentence is in difficulty, we revise.

```
When the larval stage of the parasitic worm was ex-
posed to test animals of a species that never
serves as a host, the larvae did not parasitize the
test animals.
```

Finally, look what can happen when a variety of these pronouns are scattered throughout a sentence.

```
Although they both saw the same things in their ob-
servations of embryonic development, they had dif-
ferent theories about how this came about.
```

A patient reader of the whole essay could probably figure out this sentence eventually, but its author has certainly violated rule 8 (page 8, "Never make the reader back up") in a most extreme fashion.

In short, when editing your work, read it carefully and with skepticism, checking that you have said exactly what you mean. Never make the reader guess what you have in mind. Never give the reader cause to wonder whether, in fact, you have anything in mind. Everything you write must make sense—to yourself and to the reader. As you read each sentence you have written, think: what does this sentence say? what did I mean it to say? Make each sentence work on your behalf, leading the reader easily from fact to fact, from thought to thought.

Please note that you need not be a grammarian to write correctly and clearly. With a little practice, especially if you read your work aloud, you can quickly learn to recognize a sentence in difficulty and sense how to fix it without even knowing the name of the grammatical rule that was violated.

REVISING FOR COMPLETENESS

Make sure each thought is complete. Be specific in making assertions. The following statement is much too vague:

> Many insect species have been described.

How many is "many"? After editing, the sentence might read,

> Nearly one million insect species have been described.

Similarly, the sentence

> More caterpillars chose diet <u>A</u> than diet <u>B</u> when given a choice of the two diets (Fig. 2).

would benefit from the following alteration:

> Nearly five times as many caterpillars chose diet <u>A</u> than diet <u>B</u> when given a choice of the two diets (Fig. 2).

Here is another kind of incompleteness.

> If diffusion was entirely responsible for glucose transport, then this would not have occurred.

This rears its ugly head again; the author avoids the responsibility of drawing a clear conclusion and forces the reader to back up and attempt to summarize the findings. Even the beginning of the sentence is unnecessarily vague since, it turns out, the discussion is concerned only with glucose transport in intestinal tissue. Try to make your sentences tell a more detailed story, as in this revision:

> If diffusion was entirely responsible for glucose transport into cells of the intestinal epithelium, transport would have continued when I added the inhibitors.

In the same way, "Cells exposed to copper chloride divided at normal rates" is a substantial improvement over "The copper chloride treatment was not affected."

Be especially careful to revise for completeness whenever you find that you have written *etc.,* an abbreviation for the Latin term *et cetera,* meaning "and others" or "and so forth." In writing a first draft, use *etc.* freely when you'd rather not interrupt the flow of your thoughts by thinking about exactly what "other things" you have in mind. When revising, however, replace each *etc.* with words of substance; in scientific writing, an *etc.* makes the reader suspect fuzzy thinking. You should find yourself thinking, "What, exactly, *do* I have in mind here?" If you come up with additional items for your list, add them. If you find that you have nothing to add, simply replace the *etc.* with a period and you will have produced a shorter, clearer, sentence.

Consider the following sentence and its two improvements.

ORIGINAL VERSION

> Plant growth is influenced by a variety of environmental factors, such as light intensity, nutrient availability, etc.

REVISION 1

```
Plant growth is influenced by a variety of environ-
mental factors, such as light intensity, day
length, nutrient availability, and temperature.
```

REVISION 2

```
Plant growth is influenced by such environmental
factors as light intensity, day length, nutrient
availability, and temperature.
```

In the original version, the author has dodged the responsibility of clear writing, forcing the reader to determine what is meant by *etc.* The sentence, although grammatically correct, is incomplete, waiting for the reader to fill in the missing information. The reader may justifiably wonder whether the writer knows what other factors affect plant growth. Both revised versions clearly indicate what the author had in mind. Revising for completeness often requires you to return to your notes or to the sources upon which your notes are based.

REVISING FOR CONCISENESS

Omitting unnecessary words will make your thoughts clearer and more convincing. In particular, such phrases as, "It should be noted that," "It is interesting to note that," and "The fact of the matter is that" are common in first drafts, but should be ruthlessly eliminated in preparing the second. Such verbal excess also takes less conspicuous forms. How might you shorten this next sentence?

```
Dr. Smith's research investigated the effect of
pesticides on the reproductive biology of birds.
```

Who did the work: Dr. Smith or his research? A reasonable revision would be

```
Dr. Smith investigated the effect of pesticides on
the reproductive biology of birds.
```

We have eliminated one word, and the sentence has not suffered a bit. Working on the sentence further, we can replace "the reproductive biology of birds" with "avian reproduction," achieving a net reduction of three more words.

```
Dr. Smith investigated the effect of pesticides on
avian reproduction.
```

The next example requires similar attention.

```
It was found that the shell lengths of live snails
tended to be larger for individuals collected
closer to the low tide mark (Fig. 1).
```

A good editor would eliminate the first phrase of that sentence and prune further from there. In particular, what does the author mean by "tended to be larger"? Here are two improved versions of the sentence.

```
Live snails collected near the low tide mark had
greater average shell lengths (Fig. 1).
```

```
Snails found closer to the low tide mark typically
had larger shells (Fig. 1).
```

These and most other wordy sentences suffer from one or several of four major ailments and can be restored to robust health by obeying the corresponding Four Commandments of Concise Writing.

First Commandment: Eliminate Unnecessary Prepositions

Consider this example:

```
The results indicated a role of hemal tissue in
moving nutritive substances to the gonads of the
animal.
```

Any sentence containing such a long string of prepositional phrases—"of . . . tissue," "in moving . . . substances," "to the gonads," "of the animal"—is automatically a candidate for the editor's operating table. This sentence actually contains a simple thought, buried amid a clutter of unnecessary words. After surgery, the thought emerges clearly.

```
The results indicated that hemal tissue moved nu-
trients to the animal's gonads.
```

Here is another example:

```
The cells respond to foreign proteins by rapidly di-
viding and starting to produce antibodies reactive
to the protein groups that induced their production.
```

The reader's head spins, an effect avoided by the following more concise incarnation of the same sentence:

```
In the presence of foreign proteins, the cells di-
vide rapidly and produce antibodies against those
proteins.
```

By eliminating prepositions, "Gould arrives at the conclusion that . . ." becomes "Gould concludes that. . . ." "Grazing may constitute a benefit to . . ." becomes "Grazing may benefit. . . ."

"These data appear to be in support of the hypothesis that . . ." becomes "These data appear to support the hypothesis that. . . ." And "Schooling of fish is a well documented phenomenon" becomes "Fish schooling is well documented."

Second Commandment: Avoid Weak Verbs

Formal scientific writing is often confusing—and boring—because the individual sentences contain no real action; commonly, the colorless verb *to be* is used where a more vivid verb would be more effective, as in this example:

> The fidelity of DNA replication is dependent on the fact that DNA is a double-stranded polymer held together by weak chemical interactions between the nucleotides on opposite DNA strands.

This patient suffers from Wimpy Verb Syndrome, with a slight touch of Excess Prepositional Phrase. There is *potential* action in this sentence, but it is sound asleep in the verb "is dependent." Converting to the stronger verb "depends," we read,

> The fidelity of DNA replication depends on the fact that DNA is a double-stranded polymer. . . .

But why stop there? Let's eliminate some clutter ("on the fact that") and another weak verb ("is").

> The fidelity of DNA replication depends on DNA being a double-stranded polymer. . . .

Similarly,

> Plant vascular tissues function in the transport of food through xylem and phloem.

can be enlivened by converting the phrase "function in the transport of" to the more vigorous verb "transport:"

```
Plant vascular tissues transport food through xylem
and phloem.
```

Note that by choosing a stronger verb, we have also eliminated two prepositional phrases ("in the transport of" and "of food"). Step by step, the sentence becomes shorter and clearer. As often happens during revision, fixing one problem reveals an additional problem, in this case a fundamental structural weakness that makes the reader wonder whether the student understands the relationship between "plant vascular tissues" and "xylem and phloem." Revising now for content, we might rewrite the sentence as

```
Plant vascular tissues (the xylem and phloem)
transport nutrients throughout the plant.
```

or

```
Plants transport nutrients through their vascular
tissues, the xylem and phloem.
```

Third Commandment: Do Not Overuse the Passive Voice

The passive voice is often a great enemy of concise writing, in part because the associated verbs are weak. If the subject (rats and mice, in the following example) is on the receiving end of the action, the voice is passive.

```
Rats and mice were experimented on by him.
```

If, on the other hand, the subject of a sentence ("He," in the coming example) is on the delivering end of the action, the voice is said to be active.

```
He experimented with rats and mice.
```

Note that the active sentence contains only six words, whereas its passive counterpart contains eight. In addition to creating excessively wordy sentences, the passive voice often makes the active agent anonymous, and a weaker, sometimes ambiguous sentence may result:

```
Once every month for two years, mussels were col-
lected from five intertidal sites in Barnstable
County, MA.
```

Whom should the reader contact if there is a question about where the mussels were collected? Were the mussels collected by the writer, by fellow students, by an instructor, or by a private company? Eliminating the passive voice clarifies the procedure:

```
Once every month for two years, members of the
class collected mussels from five intertidal sites
in Barnstable County, MA.
```

Similarly, "It was found that" becomes "I found," or "we found," or, perhaps, "Siegel (1986) found." Whenever it is important, or at least useful, that the reader know who the agent of the action is, and whenever the passive voice makes a sentence unnecessarily wordy, use the active voice.

Passive: Little is known of the nutritional requirements of these animals.

Active: We know little about the nutritional requirements of these animals.

Passive: The results were interpreted as indicative of. . . .

Active: The results indicated. . . .

Passive: In the present study, the food value of seven diets was compared, and the chemical composition of each diet was correlated with its nutritional value.

Active: In this study, I compared the food value of seven diets and correlated the chemical composition of each diet with its nutritional value.

Note in this last example that it is perfectly acceptable to use the pronoun I in scientific writing; switching to the active voice expresses thoughts more forcibly and clearly and often eliminates unnecessary words.

Fourth Commandment: Make the Organism the Agent of the Action

Consider this example:

```
Studies on the rat show that the activity levels
vary predictably during the day (Hatter, 1976).
```

This is not a terrible sentence, but it can be improved by moving the action from the studies ("Studies . . . show") to the organism involved, the rat:

```
Rats vary their activity levels predictably during
the day (Hatter, 1976).
```

The revised sentence is shorter, clearer, and more interesting because now an organism is *doing* something. Along the way, a prepositional phrase ("on the rat") has vanished. Alternatively, one could include the researcher in the action.

```
Hatter (1976) showed that rats vary their activity
levels predictably during the day.
```

Similarly, redirecting the action transforms

```
Increases in salinity increased larval growth rates
in Experiment I, but not in Experiment II.
```

into

```
Larvae grew faster at higher salinities in Experi-
ment I, but not in Experiment II.
```

and transforms

```
The reaction rate increased as pH was increased
from 6.0 to about 8.0, and then declined between a
pH of 8.0 and 8.5 (Figure 1).
```

to

```
Trypsin was maximally effective at pHs between
about 8.0 and 8.5 (Figure 1).
```

Note that in the original version of this last example the author redrew the graph in words: we can easily picture him or her staring at the graph and its axes while writing. In the revised version, the author makes the enzyme the agent of the action and the message comes through much more clearly.

Be a person of few words; your readers will be grateful.

REVISING FOR FLOW

A strong paragraph—indeed, a strong paper—takes the reader smoothly and inevitably from a point upstream to one downstream. Link your sentences and paragraphs using appropriate transitions so that the reader moves effortlessly and inevitably from one thought to the next, logically and unambiguously. Minimize turbulence. Always remind the reader of what has come before, and help the reader anticipate what is coming next. Consider the following example:

> Since aquatic organisms are in no danger of drying out, gas exchange can occur across the general body surface. The body walls of aquatic invertebrates are generally thin and water permeable. Terrestrial species that rely on simple diffusion of gases through unspecialized body surfaces must have some means of maintaining a moist body surface, or must have an impermeable outer body surface to prevent dehydration; gas exchange must occur through specialized, internal respiratory structures.

This example gives the reader a choppy ride indeed, and cries out for careful revision, not of the ideas themselves but of the way they are presented. In the following revision, note the effect of two important transitional expressions, *thus* and *in contrast to.* The first connects two thoughts, and the second warns the reader of an approaching shift in direction.

> Since aquatic organisms are in no danger of drying out, gas exchange can occur across the general body surface. Thus, the body walls of aquatic invertebrates are generally thin and water permeable, facilitating such gas exchange. In contrast to the simplicity of gas exchange mechanisms among aquatic species, terrestrial species that rely on simple diffusion of gases through unspecialized body surfaces must either have some means of maintaining a moist body surface, or must have an impermeable outer body covering that prevents dehy-

```
dration. If the outer body wall is impermeable to
water and gases, respiratory structures must be
specialized and internal.
```

In the first draft, the reader must struggle to find the connection between sentences. In the revised version, the writer has assisted the reader by connecting the thoughts, resulting in a more coherent paragraph.

Here is one more example of a stagnating paragraph that carries its reader nowhere:

```
     The energy needs of a resting sea otter are
three times those of terrestrial animals of compa-
rable size. The sea otter must eat about 25% of its
body weight daily. Sea otters feed at night as well
as during the day.
```

Revising for improved flow, or coherence, produces the following paragraph. Note that the writer has introduced no new ideas. The additions, here underlined, are simply clarifications that make the connections between each point explicit:

```
     The energy needs of a resting sea otter are
three times those of terrestrial animals of compa-
rable size. To support such a high metabolic rate,
the sea otter must eat about 25% of its body weight
daily. Moreover, sea otters feed continually, at
night as well as during the day.
```

The following transitional words and phrases are especially useful in linking thoughts to improve flow: *in contrast, however, although, thus, whereas, even so, nevertheless, moreover, despite, in addition to.*

Repetition and summary are also highly effective ways to link thoughts. For instance, repetition has been used to connect the first two sentences of the revised example about sea otters: "To support such a high metabolic rate" essentially repeats, in summary form, the information content of the first sentence. Repetition is a particularly effective way of linking paragraphs; in reminding the reader of what has come before, the author consolidates his or her position and then moves on. Use these and similar transitions to move the reader smoothly from the beginning of your paper to the end. Be certain that each sentence—and each paragraph—sets the stage for the one that follows, and that each sentence—and each paragraph—builds on the one that came before.

Judicious use of the semicolon can also ease the reader's journey. In particular, when the second sentence of a pair explains or clarifies something contained in the first, you may wish to combine the two sentences into one with a semicolon. Consider the following two sentences:

> This enlarged and modified bone, with its associated muscles, serves as a useful adaptation for the panda. With its "thumb," the panda can easily strip the bamboo on which it feeds.

The reader probably has to pause to consider the connection between the two sentences. Using a semicolon, the passage would read

> This enlarged and modified bone, with its associated muscles, serves as a useful adaptation for the panda; with its "thumb," the panda can easily strip the bamboo on which it feeds.

The semicolon links the two sentences and eliminates an obstruction in the reader's path. Similarly, a semicolon provides an effective connection between thoughts in the following two examples:

```
Recently we demonstrated the rapid germination of
radish seeds; nearly 80% of the seeds germinated
within three days of planting.

Recombinant DNA technology enables large-scale pro-
duction of particular gene products; specific genes
are transferred to rapidly dividing host organisms
(yeast or bacteria), which then transcribe and
translate the introduced genetic templates.
```

REVISING FOR TELEOLOGY AND ANTHROPOMORPHISM

Remember, organisms do not act or evolve with intent (pp. 11–12). Consider the following examples of teleological writing, and learn to recognize the trend in your own work:

```
Barnacles are incapable of moving from place to
place, and therefore had to evolve a specialized
food-collecting apparatus in order to survive.

Squid and most other cephalopods lost their exter-
nal shells in order to swim faster, and so better
compete with fish.

Aggression is a directed behavior that many sea
anemones exhibit to promote the survival of an in-
dividual's own genotype.
```

Revise all teleology out of your writing.

Also beware of anthropomorphizing, in which you give human characteristics to nonhuman entities, as in this example:

```
The existence of sage in the harsh climate of the
American plains results from Nature's timeless
experimentation.
```

Again, this conveys a rather fuzzy picture about how natural selection operates. The author would be on firmer ground by writing something like

```
Sage is one of the few plants capable of withstand-
ing the harsh, dry climate of the American plains.
```

REVISING FOR SPELLING ERRORS

Misspellings convey the impression of carelessness, laziness, or perhaps even stupidity. These are not advisable images to present to instructors, prospective employers, or the admissions officers of graduate or professional programs. Using a spelling-checker computer program will save you from misspelling many nontechnical words, but it won't catch such spelling errors as "is" versus "if," or "nothing" versus "noting," and it is unlikely to be of much help in screening technical terms for you. Use the computer for a "first pass," but use your own eyes for the second.

It helps to keep a list of words that you find yourself using often and consistently misspelling. *Desiccation* and *argument* were on my list for quite some time; *proceed* and *precede* are still on it. When in doubt, use a dictionary. And if you add technical terms to your computer program's dictionary, be careful to enter the correct spellings.

A few peculiarities of the English language are worth pointing out.

1. *Mucus* is a noun; as an adjective, the same slime becomes *mucous.* Thus, many marine animals produce mucus, and mucous trails are produced by many marine animals.

2. *Seawater* is always a single word. *Fresh water,* however, is two words as a noun and one word as an adjective. Thus, freshwater animals live in fresh water.
3. *Species* is both singular and plural: one species, two species. But the plural of *genus* is *genera:* one genus, two genera.

And don't forget to underline or italicize scientific names: *Littorina littorea* (the periwinkle snail), *Chrysemys picta* (the eastern painted turtle), *Taraxacum officinale* (the common dandelion), *Homo sapiens* (the only animal that writes laboratory reports).

REVISING FOR GRAMMAR AND PROPER WORD USAGE

Appendix C lists a number of books that include excellent sections on grammar and proper word usage. While on the lookout for sentence fragments, run-on sentences, faulty use of commas, faulty parallelism, incorrect agreement between subjects and verbs, and other grammatical blunders, you should also be on the lookout for violations of six especially troublesome rules of usage when revising your work.

1. *between* and *among. Between* (from *by twain*) usually refers to only two things:

    ```
    The 20 caterpillars were randomly distributed
    between the two dishes.
    ```

    Among usually refers to more than two things.

    ```
    The 20 caterpillars were randomly distributed
    among the eight dishes.
    ```

2. *which* and *that.* Most of your *which*s should be *that*s.

```
This fish, which lives at depths up to 1000 m,
experiences up to 100 atmospheres of pressure.
```

```
A fish that lives at a depth of 1000 m is ex-
posed to 100 atmospheres of pressure.
```

In the first example, "which" introduces a nondefining, or nonrestrictive clause. The introduced phrase is, in effect, an aside, adding extra information about the fish in question; the sentence would survive without it. On the other hand, the "that" of the second example introduces a defining, or restrictive clause; we are being told about a particular fish, or type of fish, one that lives at a depth of 1000 m.

Improper use of *that* and *which* can occasionally lead to ambiguity or falsehood. Consider the following sentence about the production of proteins from messenger RNA (mRNA) transcripts:

```
This difference in protein production is due to
different amounts of mRNA that translate and
produce each particular protein.
```

Here, "that" correctly introduces a restrictive clause. Which mRNA molecules? The ones coding for these particular proteins. The writer is telling us that proteins are produced in proportion to the number of mRNA molecules coding for them within the cell. Replacing "that" with "which" drastically changes the meaning of the sentence.

```
The difference in protein production is due to
different amounts of mRNA, which translates and
produces each particular protein.
```

The sentence has lost clarity because "which" now introduces a nondefining clause that should be explaining only what mRNA does, in general. In the following sentence, using the word *which* conveys information that is actually wrong:

```
In squid and other cephalopods, which lack ex-
ternal shells, locomotion is accomplished by
contracting the muscular mantle.
```

Here, the writer asserts that no cephalopods have external shells, which is not the case; some species *do* have external shells. The correct word is "that:"

```
In squid and other cephalopods that lack exter-
nal shells. . . .
```

Now the writer correctly refers specifically to those cephalopods with external shells.

As in the examples given, *which* is commonly preceded by a comma. When deciding between *which* and *that* in your own writing, read your sentence aloud. If the word doesn't need a comma before it for the sentence to make sense, the correct word is probably *that.* If you hear a pause when you read, signifying the need for a comma, the correct word is probably *which.*

3. *its* and *it's. It's* is always an abbreviated form of *it is.* If *it is* does not belong in your sentence, use the possessive pronoun *its.*

```
When treated with the chemical, the protozoan
lost its cilia and died.
```

```
It's clear that the loss of cilia was caused by
treatment with the chemical.
```

While we're at it, let's revise that last sentence to eliminate the passive voice:

```
It's clear that treatment with the chemical
caused the loss of cilia.
```

In general, contractions are not welcome in formal scientific writing. Thus, you can avoid the problem entirely by writing *it is* when appropriate:

```
It is clear that treatment with the chemical
caused the loss of cilia.
```

4. *effect* and *affect*. *Effect* as a noun means a "result" or "outcome:"

```
What is the effect of fuel oil on the feeding
behavior of sea birds?
```

Effect as a verb means "to bring about:"

```
What changes in feeding behavior will fuel oil
effect in sea birds?
```

Affect as a verb means "to influence" or "to produce an effect upon:"

```
How will the fuel oil affect the feeding behav-
ior of sea birds?
```

Used as a verb, "effect" can indeed be replaced in the preceding example by *bring about,* but not by *influence;* and "affect" can indeed by replaced by *influence,* but not by *bring about.* Even so, memorizing the definitions of the two words may be of little help in deciding which word to use in your own writing since, as verbs, *affect* and *effect* are so similar in meaning. You may be more successful in choosing the correct word by memorizing each of the examples and then comparing the memorized examples with your own sentences.

5. *i.e.* and *e.g.* These two abbreviations are not interchangeable. *I.e.* is an abbreviation for *id est,* which, in Latin, means "that is" or "that is to say." For example,

    Data on sex determination suggest that this species has only two sexual genotypes, i.e., female (XX) and male (XY).

    The embryos were undifferentiated at this stage of development; i.e., they lacked external cilia and the gut had not yet formed.

    In contrast, *e.g.* stands for *exempli gratia,* which means "for example." I will give two examples of its use:

    During the precompetent period of development, the larvae cannot be induced to metamorphose (e.g., Crisp, 1974; Bonar, 1978; Chia, 1978; Hadfield, 1978).

    However, the larvae of several butterfly species (e.g., Papilio demodocus Esper, P. eurymedon, and Pieris napi) are able to feed and grow on plants that the adults never lay eggs on.

    In the first case, the writer uses *e.g.* to indicate that what follows is only a partial listing of references supporting the statement: "for example, see these references," in other words. In the second case, the writer uses it to indicate only a partial list of butterfly species that don't lay eggs on all suitable plants.

6. *However, therefore,* and *moreover.* These words are often incorrectly used as conjunctions, as in the following examples:

    The brain of a toothed whale is larger than the human brain, however the ratio of brain to body weight is greater in humans.

    The resistance of mosquito fish (Gambusia affinis) to the pesticide DDT persisted into the next generation bred in the laboratory, therefore the resistance was probably genetically based.

    Protein synthesis in frog eggs will take place even if the nucleus is surgically removed, moreover the pattern of protein synthesis in such enucleated eggs is apparently normal.

    These examples all demonstrate the infamous comma splice, in which a comma is mistakenly used to join what are really two separate sentences. Reading aloud, you should hear the material come to a complete stop before the words "however," "therefore," and "moreover." Thus, you must replace the commas with either a semicolon or a period, as in these revisions of the first example:

    The brain of a toothed whale is larger than the human brain; however, the ratio of brain to body weight is greater in humans.

```
The brain of a toothed whale is larger than the
human brain. However, the ratio of brain to
body weight is greater in humans.
```

7. And don't forget: The data *are* . . . (see page 13).

BECOMING A GOOD EDITOR

The best way to become an effective reviser of your own writing is to become a critical reader of other people's writing. Whenever you read a newspaper, magazine, or textbook, be on the lookout for ambiguity and wordiness, and think about how the sentence or paragraph might best be rewritten. You will gradually come to recognize the same problems, and the solutions to these problems, in your own writing. But don't try to fix everything at once. Whether you are editing an early draft of your own work or a fellow student's work, be concerned first with content. Until you are convinced that the author has something to say, it makes little sense to be overly concerned with how he or she has said it, for the same reason it would make little sense to wash and wax a car that was headed for the auto salvage.

Take an especially careful look at the title and the first few paragraphs. Does the title indicate exactly what the paper or laboratory report is about? Do the title and first paragraph seem closely related? In the first one or two paragraphs, does one sentence lead logically to the next, establishing a clear direction for what follows? Can you tell from the first paragraph or two exactly what this paper, proposal, or report is about, and why the issue is of interest? Or are you reading a series of apparently unrelated facts that seem to lead nowhere or in many different directions? Does the first paragraph head in one direction, the second in another, and the third in yet another? If so, there is serious work to be done.

Second drafts commonly arise from only a small portion of the first—perhaps a few sentences buried somewhere in the last third of the original. In such a case, you must abandon most of the first draft and begin afresh, but this time you are writing from a stronger base. Always leave at least several days to make revisions, and insist that your fellow students give you drafts of their work to look over at least several days before the final piece is due.

Once the piece has a clear direction, you can revise for flow and clarity. Does each sentence make sense, and does each lead in logical fashion to the next? Does each paragraph follow logically from the previous paragraph? Does the concluding paragraph address the issue posed in the first paragraph?

If examining a laboratory report, study the Results section first. Does it conform to the requirements outlined in Chapter 3? Does the Materials and Methods section answer all procedural questions that were not addressed in the figure captions and table legends? Should some of those questions (for example, experimental temperature) be addressed in the captions and legends? Does the Introduction state a clear question and provide the background information needed to understand why that question is worth asking? Does the Discussion section interpret the data or does it simply apologize, and does the Discussion clearly address the specific issue raised in the Introduction?

Only when you can answer yes to these questions should you worry about fine-tuning the paper—editing for conciseness, completeness, grammar, and spelling.

Giving Criticism

Look at someone else's paper the same way you should look at your own, concerning yourself first with content. Avoid the temptation to smother the paper with notations about prepositional phrases and spelling errors; as discussed, it is worth commenting on such things only when you feel the piece is a draft or two away from perfection. When examining a first draft, it may be most useful to write a few paragraphs of commentary to the author and not write on the paper at all. Don't feel compelled to rewrite the paper for the author; your role is simply to point out strengths and perceived weaknesses and to offer the best advice you can about potential fixes. Here is an example of how this might be done; the student is making comments about the first draft of a research proposal written by a fellow student.

> Jim, I think you have a good idea for a project here, but it's not reflected in your introduction (or the title, but that can wait). The question you finally state in the middle of p. 4 caught me completely by surprise; at least until the bottom of p. 2 I thought you were interested in the effects of electromagnetic fields on human development, and by the end of p. 3, I wasn't sure <u>what</u> you were planning to study! On pp. 2-3 especially, I couldn't see how the indicated paragraphs (see my comments on your paper) related to the question you ended up asking. Or perhaps they <u>are</u> relevant, and you just haven't made the connections clear to me? The entire introduction seems to be in the "book report" format we discussed in class, rather than a piece of writing with a point to make (I'm having this trouble, too). The information you present is <u>interesting</u>, but a lot of it seems irrelevant. Try to make clearer connections between the paragraphs, perhaps by leaving some things out. As the Pechenik book says, "Be sure each paragraph sets the stage for the one that follows" etc.—isn't that a great book? Here is a possible reorganization plan: Introduce the concept of electromagnetic fields in the first few sentences (what they are, what produces them); then mention potential damaging effects on physiology and development (at present

it's not clear why the question is so important until one gets to p. 6!); then state your question and note why urchins are especially good animals to study. Will that work?

Also in the Introduction, I would expand the paragraph on gene expression effects; discuss one or two of the key experiments in some detail, rather than just tell us the results. I think this is important, since <u>your</u> experiments are a follow-up on these.

Your experimental design seems sound, although I'm not sure the experiments really address the exact question you pose in your introduction (see my comments on the draft; probably you just need to rephrase the question?). But I didn't see any mention of a control; without the control, how will you be sure that any effects you see are due to the electromagnetic field? Also, won't your treatment raise the water temperature? If so, you will need to control for that as well.

Finally, you might want to ask Professor Avossa about this, but I think you should write for a more scientifically advanced audience. Your tone seems a bit too chatty and informal. And watch those prepositions—you use them almost as freely as I do! I enjoyed reading your paper and look forward to seeing the next draft!

Notice that this reviewer points out the strengths of the piece without overlooking the weaknesses and deals with the major problems first. Be firm but kind in your criticism; your goal is to help your colleague, not to crush his or her ego. Be especially careful to avoid sarcasm. Write a page of constructive criticism that you would feel comfortable receiving.

Receiving Criticism

Be pleased to receive suggestions for improving your work. A colleague who returns your paper with only a smile and a pat on the back does you no favors. It is good to receive *some* positive feedback, of course, but what you are really hoping for is constructive criticism. On the other hand, don't feel you must accept every suggestion offered. Examine each one honestly and with distance, and decide for yourself if the reader is on target or not; these reviews of your work are advisory only, giving you a chance to see how another person interprets what you have written. Sometimes the reader will misinterpret your writing, and you may therefore disagree with the specific criticisms and suggestions leveled at you; however, if something was unclear to one reader, it may be equally unclear to others. Try to figure out where the reader went astray, and modify your writing to prevent future readers from following the same path.

It is hard to read criticism of your writing without feeling defensive, but learning to value those comments puts you firmly on the path to becoming a more effective writer. After all, you *want* to communicate; if you are not communicating well, you need to know it, and you need to know why.

Fine-Tuning

Once the writing is blessed with a clear direction and solid logic, it is time to make one or two final passes to see that each sentence is doing its job in the clearest, most concise fashion. As a first step in developing your ability to fine-tune writing, read the following 25 sentences and try to verbalize the ailment afflicting each one. Then revise those sentences that need help. Pencil your suggested changes directly onto the sentences, using the guide to proofreader's notation presented in Table 10.1 and the following example.

EXAMPLE

Hermaphoditism is Commonly encountered among invertebrates. For example, the young East Coast oyster, Crassostrea virginica, matures as a male, later decomes a female and may chenge sex every few years there after. sequential hermaphrodites generally change sex only once, and usually change from male to female. In contraast to species than change sex as they age, many invertebrates are simultaneous hermaprhodites. Self-fertilization is rare among simultaneous hermaphrodites it can occur, as in the tapeworms.

It is wise when editing someone else's work to use a different color pen or pencil to be sure the reader will see suggested changes.

Sentences in Need of Revision

1. To perform this experiment there had to be a low tide. We conducted the study at Blissful Beach on September 23, 1991, at 2:30 PM.
2. In *Chlamydomonas reinhardi,* a single-celled green alga, there are two matine types, + and −. The + and − cells mate with each other when starved of nitrogen and form a zygote.
3. Protruding form this carapace is the head, bearing a large pair of second antennae.

4. The order in which we think of things to write down is rarely the order we use when we explain what we did to a reader.
5. The purpose of Professor Wilson's book is the examination of questions of evolutionary significance.
6. Swimming in fish has been carefully studied in only a few species.
7. One example of this capacity is observed in the phenomenon of encystment exhibited by many fresh water and parasitic species.
8. In a sense, then, the typical protozoan may be regarded as being a single-celled organism.
9. An estuary is a body of water nearly surrounded by land whose salinity is influenced by freshwater drainage.
10. The résumé presents a summary of your educational background, research experience and goals.
11. In textbooks and many lectures, you are being presented with facts and interpretations.
12. The human genome contains at least 50,000 genes, however there is enough DNA in the genome to form nearly 2×10^6 genes.
13. It should be noted that analyses were done to determine whether the caterpillars chose the different diets at random.
14. These experiments were conducted to test whether the condition of the biological films on the substratum surface triggered settlement of the larvae.

15. Various species of sea anemones live throughout the world.
16. This data clearly demonstrates that growth rates vary with temperature.
17. Hibernating mammals mate early in the spring so that their offspring can reach adulthood before the beginning of the next winter.
18. This study pertains to the investigation of the effect of this pesticide on the orientation behavior of honey bees.
19. The results reported here have lead the author to the conclusion that thirsty flies will show a positive response to all solutions, regardless of sugar concentration (see figure 2).
20. Numbers are difficult for listeners to keep track of when they are floating around in the air.
21. Those seedlings possessing a quickly growing phenotype will be selected for, whereas. . . .
22. Under a dissecting microscope, a slide with a drop of the culture was examined at 50×.
23. Measurements of respiration by the salamanders typically took one-half hour each.
24. The results suggest that some local enhancement of pathogen specific antibody production at the infection site exists.
25. Usually it has been found that higher temperatures (30°C) have resulted in the production of females, while lower temperatures (22–27°C) have resulted in the production of males. (e.g., Bull, 1980; Mrosousky, 1982)

TABLE 10.1. PROOFREADER'S SYMBOLS USED IN REVISING COPY

Problem	Symbol	Example
1. Word has been omitted	∧ caret	Study describes ∧ effect (the)
2. Letter has been omitted	∧ caret	that b∧ok (o)
3. Letters are transposed	∼	form the sea
4. Words are transposed	∼	was only exposed
5. Letter should be capitalized	≡ (three short underlines)	these data
6. Letter should be lowercase	/ (slash)	These Data
7. Word should be in italics	___ (underline once)	Homo sapiens
8. Words are run together	| (draw vertical line in between)	edit|carefully
9. Word should be deleted	___ (draw line through)	the ~~nice~~ data
10. Space should not have been left	(sideways parentheses)	the e nd
11. Wrong letter	/ (draw line through and add correct letter above)	hemale (f)

TABLE 10.1 (Continued)

Problem	Symbol	Example
12. Wrong word	________ (draw line through and add correct word above)	*These* ~~This~~ data
13. Need to begin a new paragraph	¶ (paragraph symbol)	¶ female. In contrast
14. Restore original	(STET)	(STET) the ~~energy~~ needs

There are several ways to improve each of these sentences. For reference, my revisions are shown in Appendixes D and E, but you should make your own modifications before looking at mine. Be sure that you can identify the problem suffered by each original sentence, that you understand how that problem was solved by my revision, and that your revision also solves the problem (and does not introduce any new difficulties).

11
Answering Essay Questions

Answering essay questions on examinations differs from the other forms of scientific writing already discussed in only two respects: the essay examination must be completed within a short time, usually from 15 to 50 minutes, and you no longer have a choice in the subject of the essay. A winning answer to an essay question will follow all the guidelines outlined in Chapter 1. Your performance on essay questions can be strengthened by keeping in mind a few additional points.

1. **Read the question carefully before writing anything.** You must answer the question posed, not the question you would have preferred to see on the examination. In particular, note whether the question asks you to list, discuss, or compare. A list will not satisfy the requirements of a discussion or comparison. A request for a list tests to see whether you know all components of the answer; a request for a discussion additionally examines your understanding of the interrelationships among these components.

   Consider this list of the characteristics of a Big Mac® and a Whopper®, based on a study conducted in Somerville, Massachusetts, in October of 1996:

BIG MAC	**WHOPPER**
2 beef patties	1 beef patty
patties 3.25″ diameter	patty 3.75″–4″ diameter
fried beef	broiled beef
3-part bun (3 slices)	2-part bun (top and bottom)
sesame seeds on top bun	sesame seeds on top bun
slice of pickle	slice of pickle
chopped onion	slices of onion
slice of cheese	2–3 slices of tomato
lettuce	ketchup
sauce	mayonnaise
$2.15	$2.09
surrounded by a cardboard ring and wrapped in paper	packed in cardboard box

Suppose you are asked to write an essay presenting the features of both items. Your essay might look like this:

The Big Mac consists of two patties of fried ground beef, each patty approximately 3.25 inches in diameter, with lettuce, chopped onion, sliced pickle, a slice of cheese, some reddish sauce, and a three-part bun, with the two patties separated from each other by one of the slices of bun. The top slice of the bun is covered with sesame seeds. The Big Mac sells for $2.15 and is served in a paper wrapper, with a cardboard ring inside to hold the sandwich together.

The Whopper consists of one slice of broiled ground beef (approximately 4 inches in diameter), with mayonnaise, ketchup, several slices each of tomato, pickle, and onion, and a two-part bun, with the upper half of the bun covered with sesame seeds. The Whopper sells for $2.09 and is served in a cardboard box.

If you are asked to compare, or to compare and contrast, the two products, your essay must be written differently:

Both the Big Mac and the Whopper contain ground beef and are served on buns. The two hamburgers differ, however, with regard to the way the meat is cooked, the way the meat and bread are distributed within the hamburger, the nature of accompanying condiments, and how the sandwiches are served.

The meat in the Big Mac is fried, and each sandwich contains two patties, each approximately 3.25 inches in diameter and separated from the second patty by a slice of bun. In contrast, the meat in the Whopper is broiled, and each sandwich contains a single, larger patty, approximately 3.75—4 inches in diameter. The top bun of both sandwiches is dotted with sesame seeds. Both the Big Mac and the Whopper contain lettuce, onion, and slices of pickle. The Big Mac, however, contains chopped onion, whereas the onion in the Whopper is sliced. Moreover, the Big Mac has a slice of cheese, which is absent from the Whopper. On the other hand, the

```
Whopper comes with slices of tomato, which are
absent from the Big Mac. Both sandwiches contain
a sauce: ketchup and mayonnaise in the Whopper
and a premixed sauce in the Big Mac. The Big Mac,
at $2.15, costs 6 cents more than the Whopper.
```

If you are asked for a comparison and respond with a list, you will probably lose points, not because your instructor is being picky but because you have failed to demonstrate your understanding of the relationship between the characteristics of the two products. It is not the instructor's job to guess at what you understand; it is your job to demonstrate what you know to the instructor. Note that the facts included are the same in the two essays. The difference lies in the way the facts are presented.

If asked for a list, give a list; this response requires less time than a discussion, giving you more time to complete the rest of the examination. When asked for a discussion, discuss: present the facts and support them with specific examples. When asked for a comparison, you will generally discuss similarities and differences, but the word *compare* can also mean that you should consider only similarities. Often an instructor will ask you to compare and contrast, avoiding any such ambiguity. If you have any doubts about what is required, ask your instructor during the examination.

2. **Present all relevant facts.** Although there are many ways to answer an essay question correctly, your instructor will undoubtedly have in mind a series of facts that he or she would like to see included in your essay. That is, the ideal answer to a particular question will contain a finite number of components; the way

you deal with each of these components is up to you, but each of the components should be considered in your answer.

Before you begin to write your essay, then, list all components of the ideal answer, drawing both from lecture material and from any readings you were assigned. For example, suppose you are asked the following question:

```
Discuss the influence of physical and biological factors on the distribution of plants in a forest.
```

What components will the perfect answer to this question contain? Begin by making a list of all relevant factors as they occur to you—don't worry about the order in which you jot these factors down.

PHYSICAL	**BIOLOGICAL**
amount of rainfall	competition with other plants
annual temperature range	predation by herbivores
light intensity	
hours of light per day	
type of soil	
pesticide use	
nutrient availability	

This list is not your answer to the essay question; it is an organizing vehicle intended for your use alone. Feel free to abbreviate, especially if pressed for time ("nutr. avail.," "pred. by herbs"), but be certain you won't misunderstand your own notes while writing the essay.

In preparing to write your answer to the essay question, arrange the elements of your list in some logical order, perhaps from most to least important or so that related elements are considered together; this grouping and ordering is most quickly done by simply numbering the items in your list in the order that you decide to consider them. You have now outlined your answer; the most difficult part of the ordeal is finished.

Incorporate into your essay each of the ordered components in your list. Avoid spending all of your time discussing a few of these components to the exclusion of the others. If you discuss only four of the eight relevant issues, your instructor will be forced to assume you don't realize that the other issues are also relevant to the question posed. Show your instructor you know all the elements of a complete answer to the question.

3. **Stick to the facts.** An examination essay is not an exercise in creative writing and is not the place for you to express personal, unsubstantiated opinion. As with any other type of examination question, your instructor wishes to discover what you have learned and what you understand. Focus, therefore, on the facts, and, as with all other forms of scientific writing, support all statements of fact or opinion with evidence or example. You may wish to suggest a hypothesis as part of your essay; if so, be sure to include the evidence upon which your hypothesis is based.

4. **Keep the question in mind as you write.** Don't include superfluous information. If what you write is irrelevant to the question posed, you probably won't get additional credit for your answer, and you will most likely annoy your instructor. If what you write is not

only irrelevant but also wrong, you will probably lose points. By letting yourself wander off on tangents, you will usually gain nothing, possibly lose points, and probably lose your instructor's good will; certainly, you will waste time that might more profitably be applied elsewhere on the examination. Listing the components of your answer before you write your essay will help keep you on track.

12
Writing a Poster Presentation

Most biologists go to at least one scientific meeting each year to share their research progress with others in related fields. For many years, the standard format has been a series of 10- to 15-minute individual presentations, followed by an additional 5 minutes for questions. Because the number of sessions running concurrently in different meeting rooms has increased dramatically in recent years, creating a dilemma for listeners who must decide which of three or more equally interesting talks to attend, oral presentations are giving way to poster presentations. In a poster session, displays (called posters) containing both text and data are lined up in rows, like billboards, for all to see. Each poster represents the research of one person or research team. Each group of posters is usually displayed for only a few hours, or perhaps for one afternoon or evening, and 40 or more posters may be on display at any one time, each competing for the attention of attendees at the meeting.

Compared with oral presentations, poster presentations have the advantage that many biologists can be "talking" about their research simultaneously in a single room, and "listeners," as they stroll about the room browsing among the many posters, can have detailed conversations with the authors of those posters that they find especially interesting. The disadvantage of poster presentations is that the "speaker" no longer has a captive audience: poster sessions are like flea markets, complete with all the noise and crowds. To be successful in "selling" your information, you must plan carefully to create a display that captures the attention of browsers and then leads them through an especially clear, logical, and interesting presentation of the research; otherwise, much of your potential audience will simply pass you by, lured elsewhere by another's more compelling presentation.

How do you create a poster that people will want to stop at and read, and from which even the casual reader will take away something of substance? You need to plan a two-pronged attack:

(1) limit the amount of information you present, and

(2) arrange the information advantageously.

All too often, posters display what is essentially a full scientific manuscript—complete with formal Introduction, Materials and Methods, Results, and Discussion sections—enlarged and hung up for view, page by page. This is not a good way to attract a sizable audience for your work. It is simply not reasonable to expect people to read through 40 or 50 complete research papers during the hour or so they may spend at a particular poster session.

To be effective, your poster presentation should be streamlined to its essential findings. An effective poster does not include the same amount of detail that you would include in a formal publication or even in a talk. Your poster should be designed to inform people both within and (largely) outside your immediate field about what you have done and what you have found, and it should provide a basis for discussion with those who wish to find out more about your work. It should highlight the major questions asked, the major results obtained, and the major conclusions drawn, and it should contain the least possible amount of text.

As an example of how to construct a successful poster, let's create one based on a paper published by Richard K. Zimmer-Faust and Mario N. Tamburri in 1994, in the journal *Limnology and Oceanography* (Volume 39: 1075–1087). Normally, one gives a poster or oral presentation before publishing the work; the interested reader will profit from comparing the published paper with the poster presentation that follows. The paper reports a series of experiments defining the chemical cue that causes the swimming, microscopic larval stages of oysters to stop swimming and settle to the bottom in preparation for metamorphosing to the more familiar immobile (and highly edible) juvenile stage.

Your key weapon in attracting an audience is your title, which appears in large letters at the top of your poster. The title of the published paper, "Chemical identity and ecological implications of a waterborne, larval settlement cue," contains too little specific information to be completely compelling as a poster title; moreover, the passer-by who reads only the title leaves with nothing of substance. The poster will attract more attention and convey more information with a more revealing title, such as "Oyster larvae settle in response to arginine-containing peptides." This title indicates clearly both the question that was addressed and the key finding of the study.

The rest of the poster should focus on the Results. Only the most important results should be presented: the published paper contains seven figures and four tables; our poster will display only three of the figures, and none of the tables. To make it as easy as possible for viewers to extract the essential information we wish to convey, we want each of these figures to be self-sufficient. Each should have clearly labeled axes, contain definitions of any symbols used, and be accompanied with clear indications of the specific question being addressed and the major results found. These goals are easily accomplished, as we will see momentarily.

The bigger difficulty in achieving a completely self-sufficient figure is in explaining how the experiment was performed or how the observations were made. Don't include a detailed, formal Materials and Methods section. Instead, for each figure either (1) list the major steps taken, in numerical order, or (2) present a flowchart summary of the steps taken. You may wish to have a more detailed description of the methodology available as a one-page handout that particularly interested biologists may take with them, but the poster itself should not be cluttered with such detail. If you do accompany your poster with handouts, be sure the handout includes your name, mailing address, e-mail address, and poster title.

LAYOUT OF THE POSTER

Figure 12.1 shows a possible layout for the poster just described. Notice that it is divided into three major sections, each highlighting one key issue, and that each section is clearly separated from the other sections by substantial space. The illustrated layout makes it easy for readers to follow the logic of the presentation by scanning left to right, section by section—there is never any question of where to look next—and makes it difficult for even a casual reader to miss the point of what they are looking at.

Let's fill in the entire top section of the poster, entitled "Seawater Conditioned By Adult Oysters Stimulates Larval Settlement." Our methods section might look like this.

Methods:

1. Incubate eight adult oysters (*Crassostrea virginica*) in 16 liters of artificial seawater for 2 hours.
2. Adjust pH of oyster-conditioned seawater and control seawater to 8.0; adjust salinity to 25‰.
3. Separate oyster-conditioned seawater and control seawater samples into three molecular size fractions by dialysis.
4. Expose oyster larvae to both solutions.
5. Videotape larval behavior; determine number of individuals settling on bottom of containers by end of three minutes.

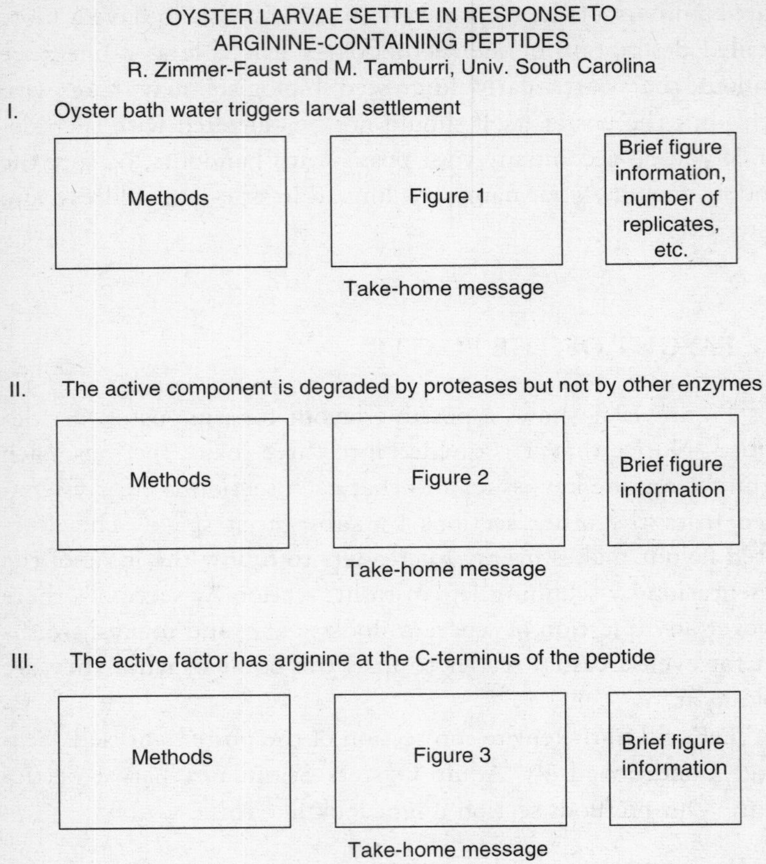

OYSTER LARVAE SETTLE IN RESPONSE TO
ARGININE-CONTAINING PEPTIDES
R. Zimmer-Faust and M. Tamburri, Univ. South Carolina

I. Oyster bath water triggers larval settlement

Methods	Figure 1	Brief figure information, number of replicates, etc.

Take-home message

II. The active component is degraded by proteases but not by other enzymes

Methods	Figure 2	Brief figure information

Take-home message

III. The active factor has arginine at the C-terminus of the peptide

Methods	Figure 3	Brief figure information

Take-home message

Figure 12.1
General layout of the poster. The goal is to make it easy for readers to see what was done and what was discovered.

Here is the same Methods information presented in flow-chart format:

Methods

Incubate eight adult oysters in 16 l of artificial seawater

↓ 2 hours

Adjust pH to 8.0. salinity to 25‰

↓

Distribute conditioned water into a glass dish

↓

Distribute control seawater into another glass dish

↓

Add sixty 20-day-old oyster larvae to each dish

↓

Videotape for three minutes

↓

Assess numbers of larvae settling to bottom in control and adult-conditioned seawater.

↓

Repeat 7 more times, using another 120 larvae per test.

The accompanying figure is shown as Figure 12.2. Each section of the poster will contain a separate Methods section and accompanying graph, following the format just presented.

MAKING THE POSTER

Well in advance of the meeting you will be told the dimensions of your display area—typically 4 feet high by 6 feet wide. Your entire display must fit within the designated area. The title of your poster should be readable from 15 to 20 feet away, so plan on using letters about 1.5 inches tall. Names of all authors and the institution they are from can be slightly smaller. The rest of your poster should be readable from about 3 to 4 feet away, so text size should be about 3/8 inch high. The easiest way to manipulate text size is by using an enlarging copying machine.

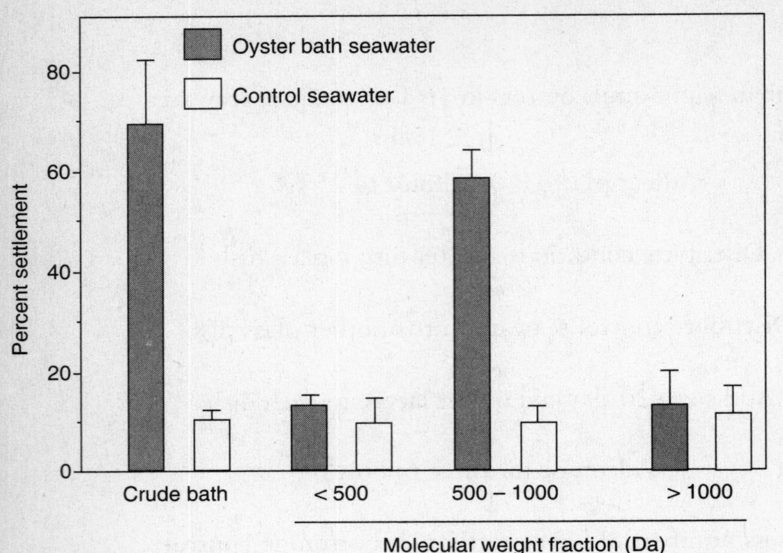

Figure 12.2
The figure and take-home message for the first section of the poster shown in Figure 12.1.

Mount the individual items of your poster on colored paper or posterboard. Use a single background color for the entire poster to unify the presentation; you may wish to use different shades of that color to better distinguish the different sections of the poster. Choose a color that provides good contrast without being jarring or distracting; brown or blue are good choices. Thumbtacks and tape, for attaching the components of your poster to the display board, are generally provided at the meetings.

When registering to present a poster or oral presentation at a meeting you are generally required to submit an Abstract of your work (see pp. 118-119), and you may be required to include the Abstract in the upper left-hand corner of your poster. If so, be sure to leave room for the Abstract in planning your layout.

Creating a successful poster takes considerable planning. But it is well worth the time and effort required: not only will you have a more productive and enjoyable meeting, you will also return from the meeting with something that can be displayed in your Biology department for students and faculty to read at their leisure.

Appendix A
Means, Variances, Standard Deviations, Standard Errors, and 95% Confidence Intervals

Suppose you have two samples of three rats each. The rat tail lengths in Samples *A* and *B* are

A = 7.0, 7.0, 7.0 cm

B = 3.6, 14.1, 3.3 cm

Both samples have the same mean value (7.0 cm), but the tails in sample *A* are much less variable in size than those in sample *B*. Simply listing the mean value, then, omits an important component of the story contained in your data.

The *variance* (σ^2) about the mean gives an indication of how variable your data are from one observation to the next. If you have access to a statistical calculator, just push the right buttons and you're almost done. If not, make your calculations using this formula.

$$\sigma^2 = \frac{\sum_{i=1}^{N} (X_i - \bar{X})^2}{N - 1}$$

N is the number of observations made, X_i is the value of the ith observation, and $\overline{X}$ is the mean value of all the observations made in a sample.

Σ is the symbol for summation. In this case, you are to sum the squared differences of each individual measurement from the mean of all the measurements. As an example, suppose you have the following data points:

5 cm
4
4
6 N=5
5

$$\overline{X} = \frac{\sum_{i=1}^{N}}{N} = \frac{24}{5} = 4.8 \text{ cm}$$

$$\sigma^2 = \frac{(5 - 4.8)^2 + (4 - 4.8)^2 + (4 - 4.8)^2 + (6 - 4.8)^2 + (5 - 4.8)^2}{4}$$

$$= 0.7$$

All you are doing is seeing how far each observation is from the mean value obtained and adding all these variations together. The squaring is done simply to eliminate minus signs so that you have only positive numbers to work with. Clearly, 100 measurements should give you a more accurate estimation of the true mean tail length than only 10 measurements, and, if you had the time and the patience, 1000 measurements would be better still. We thus divide the sum of the individual variations by a factor related to the number of observations made. Increasing the sample size will reduce the extent of experimental uncertainty. Variance, then, is a measure of the amount of confidence we can have in our measurements. The smallest possible variance is zero (all samples were identical); there is no upper limit to the potential size of the variance.

To calculate the standard deviation (SD), simply take the square root of the variance.

To calculate the standard error of the mean (SEM), simply divide the standard deviation by the square root of N.

Unlike standard deviations and standard errors, the related "95% confidence interval" has inherent meaning. If you were to repeat an experiment 100 times and calculate a mean result for each one, you can expect 95 of those means to fall within the calculated confidence interval. For samples sizes larger than about 15, the 95% confidence interval is approximately twice the standard error of the mean.

Appendix B
Commonly Used Abbreviations

	Abbreviation	Example
Length		
meter	m	3 m
centimeter (10^{-2} meter)	cm	15 cm
millimeter (10^{-3} meter)	mm	4.5 mm
micrometer (10^{-6} meter)	µm	5 µm
Weight		
gram	g	10 g
kilogram (10^{3} grams)	kg	15 kg
milligram (10^{-3} gram)	mg	16 mg
microgram (10^{-6} gram)	µm	4 µm
nanogram (10^{-9} gram)	ng	8 ng
Volume		
liter	l	3 l
milliliter (10^{-3} liter)	ml	37 ml
microliter (10^{-6} liter)	µl	13 µl
Time		
days	d	2 d
hours	h, hr	48 h, or 48 hr
seconds	s, sec	60 s, or 60 sec

Concentration		
milliosmoles/liter	mOsm/l	650 mOsm/l
molar	M	a 0.3 M solution
salinity (parts per thousand)	‰ S, ppt	31‰ S seawater, or 31 ppt
parts per million	ppm	0.2 ppm copper
parts per billion	ppb	200 ppb copper
Statistics		
mean	$\overline{X}$	$\overline{X}$ = 27.2 g/individual
standard deviation	SD	SD = 0.8
standard error	SE	SE = 0.3
sample size	N	N = 16
Other		
photoperiod (h light: h dark)	L:D	10L:14D
one species	sp.	*Crepidula* sp.
two or more species	spp.	*Crepidula* spp.
approximately	c., ≈	c. 25°C, or ≈25°C

Appendix C
Suggested References for Further Reading

GENERAL BOOKS ABOUT WRITING

Barnet, S., and M. Stubbs. 1995. *Practical Guide to Writing,* 7th ed. New York: HarperCollins College Publishers.

Hall, D., and S. Birkerts. 1991. *Writing Well,* 7th ed. New York: HarperCollins College Publishers, Inc.

Strunk, W., Jr., and E. B. White. 1979. *The Elements of Style,* 3rd ed. New York: The Macmillan Co.

BOOKS AND ARTICLES ABOUT SCIENTIFIC WRITING

Day, R. A. 1994. *How to Write and Publish a Scientific Paper,* 4th ed. Phoenix: Oryx Press.

Day, R. A. 1992. *Scientific English: A Guide for Scientists and Other Professionals.* Phoenix: Oryx Press.

Gopen, G. D. and J. A. Swan, 1990. "The Science of Scientific Writing." *American Scientist* 78: 550–558.

King, L. S. 1978. *Why Not Say It Clearly? A Guide to Scientific Writing.* Boston: Little, Brown and Co.

O'Connor, M. 1991. *Writing Successfully in Science.* New York: Chapman & Hall.

Wilkinson, A. M. 1991. *The Scientist's Handbook for Writing Papers and Dissertations.* Englewood Cliffs, NJ: Prentice-Hall.

Zinsser, W. 1995. *On Writing Well: An Informal Guide to Writing Nonfiction,* 4th ed. New York: HarperCollins Publishers, Inc.

TECHNICAL GUIDE FOR BIOLOGY WRITERS

CBE Style Manual Committee, Council of Biology Editors. 1994. *Scientific Style: The CBE Manual for Authors, Editors, and Publishers,* 6th ed. New York: Cambridge University Press.

Davis, E. B., and D. Schmidt. 1995. *Using the Biological Literature,* 2nd ed. New York: Marcel Dekker, Inc.

ADVICE ON CONSTRUCTING EFFECTIVE GRAPHS

Cleveland, W. S. 1985. *The Elements of Graphing Data.* Monterey, CA: Wadsworth Advanced Books and Software.

Tufte, E. R. 1990. *The Visual Display of Quantitative Information.* Cheshire, CT: Graphics Press.

BOOKS ABOUT WRITING FOR A GENERAL AUDIENCE

Gastel, B. 1983. *Presenting Science to the Public.* Philadelphia: ISI Press.

Nelkin, D. 1987. *Selling Science: How the Press Covers Science and Technology.* New York: W. H. Freeman and Co.

Appendix D
Revised Sample Sentences

1. ~~To perform this experiment there had to be a low tide.~~ We conducted the study at Blissful Beach [at low tide] on September 23, 1991. at ~~2:30 PM.~~
2. In *Chlamydomonas reinhardi*, a single-celled green alga,there are two mating types, + and -. The + and - cells mate with each other when starved of nitrogen and form a zygote.
3. Protruding form this carapace is the head, bearing a large mpair of second antennae.
4. The order in which we think of things to write down is rarely the order ~~we~~ use when we explain [ing] what we did to a reader.
5. ~~The purpose of~~ Professor Wilson's book ~~is the~~ examin[es]~~ation~~ ~~of~~ questions of evolutionary significance.
6. [The mechanics of] Swimming ~~in fish~~ ha[ve]s been carefully studied [for] in only a few [fish] species.
7. One example of this cap[a]city is ~~observed in~~ the ~~phenomenon of~~ encystment exhibited by many fresh water and parasitic species.
8. In a sense, then, the typical protozoan [is] ~~may be regarded as~~ being a single-celled organism.
9. An estuary is a body of water nearly surrounded by land [and] whose salinity is is influenced by freshwater drainage.
10. The résumé ~~presents a~~ summar[izes]y ~~of~~ your educational background, research experience, and goals.
11. ~~In~~ textbooks and many lectures, ~~you are being~~ presen[t] [you] with facts and interpretations.
12. The human genome ccntains 50,000 genes; however, there is enough DNA in the genome to form nearly 2 x 10^6 genes

13. ~~It should be noted that~~ analyses ~~were done~~ to determine whether the caterpillars chose the different diets at random.
14. These experiments ~~were conducted to~~ test whether the ~~condition of~~ the biological films ~~on the substratum surface~~ triggered settlement ~~of the larvae.~~
15. ~~Various species of sea anemones live throughout the world.~~

    (Sentence deleted for lack of content.)
16. This data clearly demonstrates that growth rates vary with temperature.
17. Hibernating mammals mate early in the spring ~~so that~~ their offspring ~~can~~ reach adulthood before the beginning of the next winter.
18. This study ~~pertains to the investigation of~~ the effect of this pesticide on the orientation behavoir of honey bees.
19. ~~The results reported here have lead the author to the conclusion that~~ thirsty flies ~~will~~ show a positive response to all solutions, regardless of sugar concentration (~~see~~ figure 2).
20. Numbers are difficult for listeners to keep track ~~of when they are~~ floating around in the air.
21. Those ~~seedlings possessing a quickly growing phenotype~~ will be selected for, whereas.
22. ~~Under~~ a dissecting microscope, a slide with a drop of the culture was examined at 50x.
23. Measurements of respiration ~~by the salamanders~~ typically took one-half hour each.
24. ~~The~~ results suggest that ~~some local enhancement of~~ pathogen-specific antibody produc~~tion~~ at the infection site ~~exists~~.
25. ~~Usually it has been found that higher temperatures (30°C) have resulted in the production of females, while lower temperatures (22–27°C) have resulted in the production of males. (e.g., Bull, 1980; Mrosousky. 1982)~~

    The turtles are typically born female when embryos are incubated at 30°C, and male when incubated at lower temperatures (22-27°C) (e.g., Bull, 1980; Mrosovsky, 1982).

Appendix E
The Revised Sample Sentences in Final Form

1. We conducted the study at Blissful Beach at low tide on September 23, 1991.
2. In *Chlamydomonas reinhardi,* a single-celled green alga, there are two mating types, + and –. When starved of nitrogen, the + and – cells mate with each other and form a zygote.
3. Protruding from this carapace is the head, bearing a pair of large second antennae.
4. The order in which we think of things to write down is rarely the order we use when explaining to a reader what we did.
5. Professor Wilson's book examines questions of evolutionary significance.
6. The mechanics of swimming have been carefully studied for only a few fish species.
7. One example of this capacity is the encystment exhibited by many freshwater and parasitic species.
8. In a sense, then, the typical protozoan is a single-celled organism.
9. An estuary is a body of water nearly surrounded by land and whose salinity is influenced by freshwater drainage.
10. The résumé summarizes your educational background, research experience, and goals.

11. Textbooks and many lectures present you with facts and interpretations.
12. The human genome contains at least 50,000 genes; however, there is enough DNA in the genome to form nearly 2×10^6 genes.
13. The data were analyzed to determine whether the caterpillars chose the different diets at random.
14. These experiments tested whether the biological surface films triggered larval settlement.
15. (Sentence deleted for lack of content.)
16. These data clearly demonstrate that growth rates vary with temperature.
17. Hibernating mammals mate early in the spring. As a consequence, their offspring reach adulthood before the beginning of the next winter.
18. This study describes the effect of this pesticide on the orientation behavior of honey bees.
19. Thirsty flies apparently show a positive response to all solutions, regardless of sugar concentration (Fig. 2).
20. Numbers floating around in the air are difficult for listeners to keep track of.
21. Those seedlings genetically programmed for faster growth will be selected for, whereas . . . .
22. A slide with a drop of the culture was examined at 50X using a dissecting microscope.
23. Measurements of salamander respiration typically took one-half hour each.
24. Results suggest that pathogen-specific antibodies are produced at the infection site, and thus are enhanced locally.
25. The turtles are typically born female when embryos are incubated at 30°C, and male when incubated at lower temperatures (22–27°C) (*e.g.*, Bull, 1980; Mrosousky, 1982).

Appendix F
Some Computer Software for the Biological Sciences

These are just some of the most popular of the many software programs available. Newly available software is reviewed at the back of each issue of the *Quarterly Review of Biology,* which is published four times each year.

GRAPHING PROGRAMS

SigmaPlot (for IBM and MAC)
Jandel Scientific
2591 Kerner Blvd.
San Rafael, CA 94901

Inplot (for IBM)
GraphPad Software
10855 Sorrento Valley Road, Suite 203
San Diego, CA 92121

Prism (for IBM)
GraphPad Software
10855 Sorrento Valley Road, Suite 203
San Diego, CA 92121

Kaleidograph (for MAC)
Synergy Software
2457 Perkiomen Ave.
Reading, PA 19606

DeltaGraph (for IBM and MAC)
Edutech Company
243 Foam Street
Monterey, CA 93940

BIBLIOGRAPHIC SOFTWARE

Papyrus (for IBM)
Research Software Design
2718 SW Kelly Street, Suite 181
Portland, OR 97201

Endnote (for IBM and MAC)
Niles and Associates Inc.
800 Jones Street
Berkeley, CA 94710

STATISTICAL PACKAGES

SYSTAT (for IBM)
SPSS
444 North Michigan Avenue
Chicago, IL 60611

Minitab (professional version, for IBM)
3081 Enterprise Drive
State College, PA 16801-3008

Minitab (student version)
Addison-Wesley Publishing Company
One Jacob Way
Reading, MA 01867

Statview (for IBM and MAC)
Abacus Concepts Inc.
1918 Bonita Avenue
Berkeley, CA 94704-1014

Index